Advances in Biology

nova
science publishers
www.novapublishers.com

Advances in Biology

Advances in Biology. Volume 4
Charles D. Grant (Editor)
2023. ISBN: 979-8-88697-999-2 (Hardcover)
2023. ISBN: 979-8-89113-074-6 (eBook)

Advances in Biology. Volume 3
Charles D. Grant (Editor)
2023. ISBN: 979-8-88697-779-0 (Hardcover)
2023. ISBN: 979-8-88697-846-9 (eBook)

Advances in Biology. Volume 1
Charles D. Grant (Editor)
2022. ISBN: 979-8-88697-201-6 (Hardcover)
2022. ISBN: 979-8-88697-218-4 (eBook)

Advances in Biology. Volume 2
Charles D. Grant (Editor)
2022. ISBN: 979-8-88697-422-5 (Hardcover)
2022. ISBN: 978-1-61761-293-0 (eBook)

More information about this series can be found at
https://novapublishers.com/product-category/series/advances-in-biology/

Charles D. Grant
Editor

Advances in Biology

Volume 5

nova
science publishers
www.novapublishers.com

Library of Congress Cataloging-in-Publication Data

ISBN: 979-8-89113-441-6
ISSN: 2833-1753

Published by Nova Science Publishers, Inc. † New York

Contents

Preface

This book is a compilation of eight chapters on the advancements in Biology. The first chapter comprehensively reviews advancements and challenges in utilizing extracellular polymers as materials with biomedical applications. The second chapter discusses the current mechanisms of tubulointerstitial fibrosis progression and the increasingly important role of the proximal tubule injury in driving inflammation and fibrosis. Chapter Three will discuss using computational analysis, in vitro experiments, and applied methods to provide a comprehensive approach to target the synthesis of polymeric substances in pathogenic bacteria. Chapter Four is a short review of such an approach with a special focus on the use of combinatorial peptide ligand libraries (CPLL), extensively applied in proteomics investigations, for the enrichment of very low-abundance proteins in view of their identification and individual quantitation. Chapter Five is a short review of an approach that will improve the sensitivity of analytical methods, with a special focus on the use of combinatorial peptide ligand libraries (CPLL), extensively applied in proteomics investigations, for the enrichment of very low-abundance proteins in view of their identification and individual quantitation. Chapter Five goes on to discuss the function of Cyclo (L-Phenylalanine-L-Proline) on the Proteome of Staphylococcus Aureus. In Chapter Six, recent findings and ongoing progress on Pep-triggered immunity are presented, along with the recent findings of Peps orthologs in other plant species. The mechanism of the influence of celestial bodies on activity and the interaction of domains (A- and B-Tubulins) with their surroundings are addressed in Chapter Seven. The last chapter highlights the naturally occurring thiolane type cyclic sulfides and their therapeutical potency particularly anticancer activity. Thiolane-type cyclic sulfides are especially synthesized by Allium species.

Chapter 1

Biomedical Applications of Extracellular Polymers

Yessica Enciso-Martínez[1], MS
F. Javier Vázquez-Armenta[2], PhD
B. Shain Zuñiga-Martínez[1], MS
A. Thalía Bernal-Mercado[3], PhD
Melvin R. Tapia-Rodríguez[1], PhD
and J. Fernando Ayala-Zavala[1,*], PhD

[1]Coordinación de Tecnología de Alimentos de Origen Vegetal, Centro de Investigación en Alimentación y Desarrollo, Hermosillo, Sonora, México
[2]Departamento de Ciencias Químico-Biológicas, Universidad de Sonora, Hermosillo, Sonora, México
[3]Departamento de Investigación y Posgrado en Alimentos, Universidad de Sonora, Hermosillo, Sonora, México

Abstract

This chapter comprehensively reviews advancements and challenges in utilizing extracellular polymers as materials with biomedical applications. These extracellular polymers can be included in various bio-ink formulations, from gelatin methacrylate-coated nanorods to hyaluronic acid-based hydrogels, which are discussed, highlighting their relevance in cardiac, bone, and cartilage tissue engineering. Cited studies reveal optimization strategies to enhance properties such as rheology, biocompatibility, and uniform cell incorporation. A key challenge in bioprinting is the selection of materials with suitable rheological

[*] Corresponding Author's Email: jayala@ciad.mx.

In: Advances in Biology. Volume 5
Editor: Charles D. Grant
ISBN: 979-8-89113-441-6

properties that are also biocompatible and allow for consistent cell incorporation. Issues concerning the vascularization and integration of printed tissues are examined, along with the need for standardization and rigorous evaluation to ensure clinical viability. Future perspectives indicate that incorporating extracellular polymers into bio-printed tissues could revolutionize regenerative medicine by enabling the precise recreation of the natural extracellular matrix, which is essential for tissue functionality and structure. However, further research is needed to overcome existing challenges and fully leverage the clinical potential of this emerging technology.

Keywords: polymeric matrices, therapeutic applications, 3D tissues, scaffold design

Introduction

The cornerstone of biomedical engineering lies in identifying and applying optimal materials for biological systems. The essential problem is searching for materials that can safely interface with biological tissues for therapeutic or diagnostic purposes (Zhu et al., 2017; Zhang et al., 2013; Jia et al., 2016). Materials are needed to promote cellular growth and encourage tissue regeneration while maintaining biocompatibility and biodegradability. The urgency for finding such materials is exacerbated by increasing diseases that necessitate tissue and organ replacement and the inherent challenges posed by donor shortages and allograft rejections. Thus, there is a critical need to explore and develop materials that can effectively fulfill these multiple complex roles.

The engineering of natural polymers, including extracellular substances, has emerged as a promising discipline with the potential to address various needs and challenges in the biomedical field (Reis et al., 2008). Extracellular polymers are macromolecules that compose the extracellular matrix in biological tissues, providing structural support and guiding cell function necessary in the architecture and functionality of organs and tissues (Bandzerewicz and Gadomska-Gajadhur, 2022). These polymers play a crucial role in regulating the cellular microenvironment, cell adhesion, differentiation, and migration, directly selecting tissue repair and regeneration in the organism (Ravichandran et al., 2012). The production and regulation of these extracellular polymers are essential for the maintenance and functionality of tissues, and any dysfunction in their synthesis or degradation can have significant implications for the health and homeostasis of the organism. Among the most prominent examples of extracellular polymers in the biomedical context are collagen, responsible for the resistance and elasticity of the skin and connective tissues; elastin, which gives elasticity to blood vessels and elastic tissues; and hyaluronic acid, which plays a crucial role in joint lubrication and wound healing. Chitosan, alginate, gellan, and cellulose have also received increasing attention for various uses in biomedical and pharmaceutical applications (Jorfi and Foster, 2015).

Extracellular polymers with biomedical applications can be obtained from natural or synthetic sources through recombinant production. For example, collagen can be obtained from natural sources such as animals or plants (Avila Rodríguez et al., 2018). The most common sources are from bovine, porcine, human, and marine organisms such as fish skin. Other sources include chicken, kangaroo tail, duck foot, and frog skin. Natural sources have the advantage of being easy to collect and inexpensive, but they can cause allergies and cause illness (Rezvani Ghomi et al., 2021a). On the other hand, synthetic sources have been developed to prevent immune responses, produce high-quality collagen, and avoid animal allergens (Kumar et al., 2014). Some commercial extracellular polymer substances, such as collagen, are obtained using recombinant technology produced through cells of mammals, yeasts, insects, and plants. However, these sources are costly and low yield is obtained, so animal sources are preferred (Rezvani Ghomi et al., 2021a). In the case of collagen, marine sources seem like a good option. Similar sources are found for other extracellular polymers.

Traditional approaches to sourcing materials for biomedical applications have largely been rooted in the use of synthetic or natural substances like polyvinyl alcohol (PVA) and hydroxyapatite (HA) (Bendtsen et al., 2017). While these materials have shown some promise, they are not without limitations. For instance, they often lack the complex biochemical cues necessary for guiding optimal cell behavior, are limited in biocompatibility, and may not have the required mechanical properties for load-bearing applications (Osmałek et al., 2014). Furthermore, their utility is sometimes compromised by uneven cell distribution, rendering them less than ideal for tissue engineering applications. Therefore, while conventional methods have contributed to the field, there is an evident need for more effective and versatile options.

Extracellular polymer engineering represents excellent potential in the biomedical field due to its notable advantages and diverse applications, such as tissue regeneration, tissue engineering, controlled drug release, and bioprinting. The polymer materials for biomedical applications must be biocompatible, not elicit an unnecessary inflammatory response, and not demonstrate any adverse immune response or cytotoxicity (Acosta et al., 2020). In this context, the importance of developing materials and techniques that can accurately mimic the complexity of extracellular polymers is highlighted to advance regenerative medicine and develop more effective and safe therapies.

One of the main advantages of extracellular polymer engineering is the ability to design biomaterials that mimic the complexity and functionality of extracellular polymers in natural tissues (Boni et al., 2018). These biomaterials can provide a microenvironment conducive to cell growth and differentiation, facilitating the regeneration of damaged or diseased tissues to improve the treatment of diseases and injuries and provide advanced solutions for various therapeutic applications (Shin et al., 2021). In addition, extracellular polymer engineering allows for incorporating growth factors and bioactive molecules into three-dimensional scaffolds, thus promoting targeted cell stimulation and tissue-specific development (Ravichandran et al., 2012). This ability to personalize biomaterials opens the door to more effective regenerative therapies and personalized treatments for patients with various medical conditions.

Extracellular polymer engineering has many therapeutic applications in the biomedical field, as stated before. One of the most promising areas is tissue regeneration, where engineered biomaterials can function as scaffolds to support the growth and development of specific tissues (Song et al., 2018). For example, the engineering of extracellular polymers has been used successfully in regenerating skin, bone, cartilage, and nerve tissue, opening new possibilities for repairing tissues damaged by injuries or degenerative diseases (Charbe et al., 2023; Sahana and Rekha, 2018). In addition, these biomaterials can be used in tissue engineering, allowing the creation of functional organs and tissues that can be used in transplants, overcoming the limitations associated with the scarcity of donors and the possibility of organ rejection (O'Brien, 2011). Another critical application of extracellular polymer engineering is the controlled release of drugs. By incorporating drugs within biomaterials, it is possible to achieve a gradual and sustained release of therapeutic agents, which can significantly improve treatment efficacy and reduce unwanted side effects (Lim et al., 2019). This controlled release approach allows for the design of more precise and targeted drug delivery systems, which is essential in treating chronic diseases or conditions requiring prolonged doses of drugs (Adepu and Ramakrishna, 2021).

Despite the promising potential of extracellular polymer engineering, it also faces technical, biological, and clinical challenges that must be overcome for its successful implementation. Proper selection of biomaterials, engineering of scaffolds with complex architectures, vascularization of imprinted tissues, and regulation of cell behavior are vital challenges that must be addressed to achieve success in this area. However, these difficulties also provide exciting opportunities for research and development in the field, with the potential to drive significant advances in regenerative medicine and cell therapy. This chapter aims to serve as a comprehensive repository of the current research on using extracellular polymers for biomedical applications. It seeks to analyze the various bioink formulations developed, their implications in tissue engineering, and the challenges in their clinical adoption. This analysis will delve into the complexities of biomaterial selection, evaluating both their current capabilities and limitations. It will offer a forward-looking perspective on the promising avenues for future research.

Biomedical Challenges: Scarcity of Suitable Biomaterials, Inefficiencies in Drug Delivery, Constraints in Disease Modeling, and Bioprinting Challenges

The critical challenges in the biomedical engineering landscape span several areas, each with unique difficulties that compound the overarching issues in tissue engineering and regenerative medicine (Figure 1). First and foremost is the scarcity of suitable biomaterials that can replicate natural tissues' complex biochemical and mechanical environment (Zhu et al., 2017). This lack stems from the need for materials that are not only biocompatible but also possess adequate mechanical strength for structural support. Current options often require a trade-off between these essential features, leaving clinicians and researchers constantly searching for more effective alternatives (Zhang et al., 2013).

Biomedicine has made remarkable advances in recent decades, revolutionizing healthcare and our understanding of human health and disease. Advances in genomics and molecular biology have paved the way for personalized medicine, allowing for tailored treatments based on individual genetic profiles (Malone et al., 2020). Furthermore, advanced imaging technologies and non-invasive diagnostic tools have revolutionized early disease detection and monitoring, improving patient outcomes and quality of life (Feng et al., 2021; Jartarkar et al., 2021). Regenerative medicine and tissue engineering have made significant progress in generating functional tissues and organs, giving patients with organ failure or debilitating injuries new hope (de Sousa Victor et al., 2020). Precision medicine, targeted therapies, and immunotherapies have improved cancer treatment, resulting in higher survival rates and fewer side effects (Martin et al., 2020). These ongoing advancements, combined with the incorporation of bioprinting, artificial intelligence, and big data analytics, continue pushing the boundaries of biomedicine, paving the way for a future of transformative healthcare solutions that promise to improve human health and longevity. However, there are still challenges that biomedicine must overcome; some examples are described in this section.

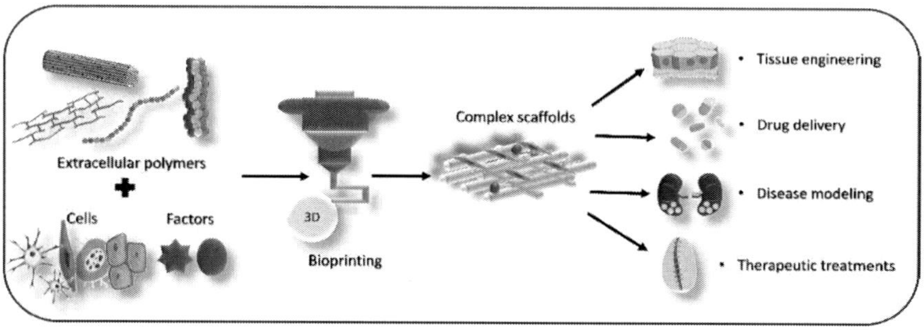

Figure 1. Facing biomedical challenges using extracellular polymers.

Humans have long wished to be able to completely repair or regenerate damaged tissues or organs and return them to their original functions (Han et al., 2020). Through organ transplantation, lost or damaged tissue can be successfully replaced, and its physiological and metabolic functions can be recovered. Many seriously ill patients have been given new lives thanks to whole organ transplants, which restored their vital functions (Charbe et al., 2023). However, conventional transplant surgeries with both autologous and allogeneic grafts have severe limitations, including the risk of organ rejection, immunogenicity or disease transmission, the scarcity of available and suitable donors' tissues, and the need for regulatory approvals (Naderi et al., 2011). Tissue engineering provides a solution to these problems by replacing and regenerating damaged tissues and organs by combining stem cells with bioactive molecules onto a highly porous scaffold of three-dimensional (3D) biomaterials which operate as templates for tissue regeneration or to direct the creation of new tissue (O'Brien, 2011). Tissue engineering is based on the principles of biomaterial science, transplantation, cell biology, and engineering that facilitate the development of cellularized structures to treat or replace damaged tissues and restore their natural function (Naderi et al., 2011). Artificial skin and cartilage are examples of FDA-approved synthetic tissue; nonetheless, complex tissues or organs such as the heart, muscle, kidney, liver, and lung remain under study (Charbe et al., 2023).

Despite the advances and promising potential of tissue engineering, this innovative technology faces several limitations that must be addressed for its widespread clinical application. A significant burden is the complexity and limited functionality of generated tissues and organs compared to their natural counterparts. Although successes have been achieved in creating essential tissues, such as skin and cartilage, reproducing highly vascularized and complex structures, such as the heart or liver, remains a significant challenge (Naderi et al., 2011). Furthermore, the growth rate and maturation of the generated tissues are often not comparable to natural human tissues, which can affect their functionality and ability to integrate properly into the recipient's body (Dzobo et al., 2018).

One of the significant challenges in tissue engineering is the scarcity of biomaterials with the characteristics required to simulate the complex nature of the natural extracellular matrix. It explores the difficulties in finding or producing biomaterials with biocompatibility, mechanical strength, acceptable degradation rates, and bioactive characteristics necessary for specific biomedical applications (Acosta et al., 2020). Biomaterials and cell sources should be tailored to each engineered tissue or organ. A biomaterial is a material intended to interface with biological systems to evaluate, treat, augment, or replace any body tissue, organ, or function (O'Brien, 2011). In the production of scaffolds for tissue engineering, three distinct types of biomaterials are usually used: ceramics (hydroxyapatite and tri-calcium phosphate), synthetic polymers (polystyrene, poly-l-lactic acid, polyglycolic acid, and poly-dl-lactic-co-glycolic acid), and natural polymers. In addition, there is also functional classification such as they are hydrogels (Janarthanan and Noh, 2021), surface modified (Sun et al., 2020), and capable of drug delivery (Sood et al., 2021), among others. Each unique biomaterial group has distinct advantages and limitations; hence, using composite scaffolds of several phases is becoming more popular (O'Brien, 2011).

Research on natural biomaterials for tissue repair and regeneration has been dramatically accelerated due to their advantages compared to permanent prosthetic devices and synthetic biomaterials. This increased research can be attributed to their similarity in biological and mechanical functions of native extracellular matrix in body tissues. Natural biomaterials can be broadly classified: 1. Protein matrices, including collagen, gelatin, fibrin, and elastin. These materials provide a structure like the biological extracellular matrix and are excellent for promoting cell adhesion, proliferation, and differentiation; 2. Polysaccharide matrices, such as hyaluronic acid, chitosan, and alginate, are widely used in tissue engineering due to their biocompatibility and water retention capacity. These materials are ideal for cell encapsulation and controlled release of growth factors. 3. Proteoglycan matrices are made up of proteins and glycosaminoglycan chains. These biomaterials can provide structural and regenerative support for tissue engineering, especially in articular and cartilage tissue applications (Song et al., 2018). Other natural substances can include decellularized extracellular matrices obtained from biological tissues stripped of their cells, leaving the three-dimensional structure of the extracellular matrix intact and natural fibers, such as silk and fibroin, used to create three-dimensional scaffolds that mimic the form of biological tissues. These materials can be modified to improve their mechanical properties and degradation.

Natural polymers like silk fibroin, collagen, and gelatin are similar to the original extracellular matrix and have higher cell adhesion and biocompatibility than synthetic polymers (Guo et al., 2021; Shoichet, 2010). Natural biomaterials share several unique characteristics, such as biocompatibility, biodegradability, and non-toxicity. However, depending on their application, they may have certain disadvantages, for example, a) mechanical strength is limited, preventing the production of larger constructs and limiting their applicability in load-bearing locations; b) manufacturing variability; and c) Potential contaminants from proteins or polysaccharides before implantation, which can be immunogenic. Despite these drawbacks, practically every tissue and organ in the body has been successfully bioengineered *in vitro* (Brovold et al., 2018).

Another challenge biomedicine presents is drug release, where the practical and precise delivery of therapeutic agents is paramount (Zeyi Wang et al., 2021). Traditional delivery systems have some drawbacks, such as inadequate drug loading, which refers to the inability to incorporate enough of the active pharmaceutical ingredient into the drug delivery system (Meng et al., 2020). This limitation can lead to suboptimal drug concentrations at the target site, compromising the desired therapeutic effects. In addition, drug administration into body cavities (oral, rectal, vaginal) can be difficult and infeasible because they are degraded at the administration site or might cause local irritations or harm when the drug concentration is high (Jain, 2020).

Another challenge lies in burst release, which occurs when a significant amount of the drug is released from the delivery mechanism quickly after administration. Burst release can cause an early rise in drug concentration, perhaps resulting in undesirable effects or wasting the drug before it reaches its intended site of action. This is especially troublesome for medications with small therapeutic windows or those requiring sustained and controlled release over a long period (Elkomy, 2020).

Some conventional drug delivery techniques do not provide targeted specific delivery (Dang and Guan, 2020). Therapeutic medicines should ideally be directed to the afflicted area or diseased tissue while sparing healthy tissues. Without tailored administration, there is a risk of off-target effects, decreased efficacy, and higher side effects. Furthermore, accurate regulation of release kinetics is essential for tailoring pharmaceutical administration to individual therapeutic requirements (Wang et al., 2021). Some conditions require a rapid release to get immediate benefits, whereas others require a steady release over a lengthy period for long-term treatment. Inefficiencies in medication delivery can prevent the anticipated release profile from being accomplished, limiting the effectiveness of treatment.

Researchers are investigating various approaches such as nanotechnology-based drug carriers, responsive biomaterials that release drugs in response to specific cues (e.g., pH, temperature, enzymes), and controlled drug delivery platforms that integrate multiple strategies to overcome these challenges (Adepu and Ramakrishna, 2021; Meng et al., 2020; Yoo et al., 2019). Because of their potential to improve drug solubility, stability, and targeted administration, nanoparticles, liposomes, micelles, and hydrogels are among the drug delivery vehicles under significant investigation. Furthermore, advancements in biomaterial engineering, nanotechnology, and biocompatible coatings have paved the way for personalized medicine and patient-specific drug delivery systems. Tailoring drug delivery to individual patient demands could improve therapeutic outcomes while reducing adverse effects, potentially transforming illness care and pharmaceutical administration.

Some studies have shown that natural biomaterials like polymers, polysaccharides, proteins, lipids, and peptides could be used as a carrier in drug delivery systems in variable lengths from nano-sized to macro-sized in various application pathways (Lim et al., 2019). The selection of polymers or biomaterials is critical in developing a drug delivery system according to the type of formulation, physicochemical attributes, and drug release profiles. Biomaterials in drug release should be biodegradable, non-toxic, biocompatible, if possible hydrophilic, mucoadhesive, and with good mechanical strength. Controlled drug delivery employs both non-degradable and biodegradable biomaterials; however, biodegradable polymers are recommended since breakdown removes the drug delivery system from the body and aids in preventing dangerous residue accumulation. Natural polymers such as alginates, starches, dextrans, cellulose, gums, chitosan, collagens, and gelatine are used in the development of controlled release drug delivery systems, because they can respond to external physical or chemical stimuli to deliver the medication, including pH, temperature, ultrasound, electric and light stimuli (Adepu and Ramakrishna, 2021; Srivastava et al., 2016).

On the other hand, developing pharmaceuticals, for which it is necessary first to understand the diseases, is one of the significant difficulties facing the biomedical industry. Convergence of results between experiments conducted on humans and animals is quite challenging. Because medication doses, formulations, and regimens in the clinic differ from those given to the animals and because of species variations, animal studies usually fail to predict actual clinical outcomes (Ledford, 2011). To better understand disease mechanisms and drug development efforts and evaluate new therapeutic compounds, it is vital to create human surrogate models that bridge the gap between conventional rodent studies and human trials (Ebrahimkhani et al., 2014).

Disease modeling has been created to address these issues (Lancaster and Huch, 2019). Disease modeling aims to replicate and analyze pathological states or diseases in a controlled laboratory setting. Disease modeling is essential in biomedical research and therapeutic development. It aids researchers in understanding disease mechanisms, identifying biomarkers for diagnosis and prognosis, and evaluating new medications' safety and efficacy before clinical trials. Illness models also help with customized medicine, where therapies can be tailored to individual patients based on their specific illness characteristics (Davidson et al., 2015).

Even while disease modeling offers valuable insights, it is essential to understand that these models might not accurately capture the complexity of diseases in humans. The challenge of precisely recreating the complex illness microenvironment is one of the limitations of disease modeling. Extracellular matrix, blood arteries, signaling molecules, and a complex network of cells comprise highly sophisticated and dynamic human tissues. These elements interact to influence cellular activity and the onset of disease. It can be challenging to replicate this complexity in disease models, especially in two-dimensional (2D) cell cultures that might not accurately reflect the spatial structure and cellular interactions seen *in vivo* (Ebrahimkhani et al., 2014). Disease models frequently rely on easily accessible cell lines or primary cells, which may not adequately represent the diversity of cell types found in sick tissues. Obtaining or producing disease-specific cell types can be difficult, limiting the model's accuracy.

Another significant limitation in disease modeling is traditional models' lack of vascularization and immune cell infiltration (Sun et al., 2022). Many diseases have complex interactions with the immune system. Capturing these immune responses in disease models can be tricky since it necessitates incorporating immune cells and understanding the dynamic interplay between immune cells and sick tissues. Vascular networks are critical for simulating diseases with vascular involvement, such as cancer and ischemia disorders, because they give nutrition and oxygen to tissues. Incorporating 3D bioprinting or microfluidic devices into extracellular polymer-based models makes it possible to create vascular networks and examine angiogenesis and immune responses in a more realistic environment (Anthon and Valente, 2022).

Hydrogels and scaffolds made with extracellular substances can act as biomimetic matrices, providing cells with a more physiologically appropriate habitat (Koivisto et al., 2019). Hydrogels have piqued the interest of researchers because of their biocompatibility, customizable mechanical properties, and ability to retain a high-water content comparable to actual tissues (Catoira et al., 2019). These polymers enable cells to be immersed in a three-dimensional environment, enhancing cell-cell interactions and cell-matrix communication closely resembling the *in vivo* microenvironment. This allows for illness models with improved physiological relevance and a more accurate description of disease pathophysiology. Extracellular polymers can also be functionalized with bioactive molecules like growth factors, cytokines, or chemokines to mimic disease-related signaling pathways and produce disease-specific microenvironments (Lyu et al., 2020). By simulating the *in vivo* environment, these functionalized polymers can help researchers better understand disease causes and assess the efficacy of possible therapeutics. Despite significant progress, there are still obstacles in disease modeling with extracellular polymers. Replicating the deep complexity of native tissues is challenging, and researchers are constantly refining and optimizing these models to improve their relevance and predictive powers.

Bioprinting is a cutting-edge biomedical technology that involves the exact and controlled production of 2D, 3D, and 4D biological constructs with living cells, biomaterials, and bioactive components. Bioprinting is like regular 3D printing in that it uses bioinks made of living cells and biomaterials that replicate the extracellular matrix of tissues instead of synthetic materials. Bioprinting is a promising and quickly expanding technology, but it also has a few critical hurdles that must be overcome before it can be successfully clinically translated and widely applied (Gaihre et al., 2022). One of the most challenging aspects of bioprinting is guaranteeing high cell survival rates while retaining cell functionality in the printed tissues. During the bioprinting process, shear forces, pressure, and temperature variations can impact cell health and damage cell viability and performance.

Creating a functional vascular network within bio-printed tissues is critical for supplying nutrients and oxygen to cells. The absence of a circulatory supply can limit tissue thickness and cause necrosis in the construct's core. It is still challenging to print vascular systems that match the complexity and perfusion requirements of human tissues (Zhang et al., 2017). In addition, selecting appropriate biomaterials for bioinks is crucial in bioimprinting. Biomaterials must have the correct mechanical characteristics, biocompatibility, and bioactivity to support cell behavior and tissue formation. Identifying and designing biomaterials that closely resemble the original extracellular matrix of various tissues is challenging. Furthermore, regulatory approvals, cost-effectiveness, and long-term safety and efficacy studies must be considered when moving bioprinting technology from the lab to the clinic.

Extracellular polymers such as hydrogels and scaffolds of gelatin, collagen, hyaluronic acid, and alginate have been successfully used in bioprinting as bioinks or support structures (Bishop et al., 2017). These biomaterials have structural integrity, mechanical characteristics, and bioactive cues to replicate the original tissue milieu and enhance cell function and tissue formation. Biocompatible extracellular polymers promote cell survival and proliferation. They can be designed to offer an appropriate milieu for the encapsulated cells, allowing for cell adhesion, proliferation, and differentiation within the printed structures (Shin et al., 2021). Extracellular polymers can be changed to stimulate angiogenesis or act as a template for integrating endothelial cells and forming vascular channels within bioprinted constructions, enhancing tissue perfusion and functionality.

Biomedicine has several fundamental difficulties that must be addressed to develop research and improve patient outcomes. Extracellular polymers hold significant potential for addressing these challenges by providing essential biocompatibility, mechanical strength, and bioactive qualities for specific biomedical applications. These polymers can be engineered to act as drug carriers and deliverers, to be used as bio-ink in the bioprinting process by providing structural support, promoting cell adhesion and functionality, and facilitating the development of vascular networks within bio-printed constructs and creating disease models that better mimic the microenvironment and cellular interactions found in diseased tissues. The biomedical field can open new potential for cutting-edge medicines, regenerative medicine, and precision healthcare customized to individual patient needs by solving these obstacles through interdisciplinary research, technological innovation, and collaboration.

The biomedical challenges are manifold but interconnected. The scarcity of suitable biomaterials, inefficiencies in drug delivery, limitations in disease modeling, and the specialized challenges of 3D bioprinting necessitate a multidisciplinary approach. Each of these challenges will be addressed in this chapter through the lens of extracellular polymers as a novel solution.

Design and Manufacture of Three-Dimensional Extracellular Polymer Scaffolds for Tissue Engineering and Regenerative Medicine

Tissue engineering is a multidisciplinary field that combines principles of biology, engineering, and materials science to create functional tissue substitutes for regenerative medicine applications. It involves developing and implementing biomaterials, cells, and biochemical cues to engineer constructs that mimic the structure and function of native tissues. Tissue engineering strategies promote cellular adhesion, proliferation, and differentiation within three-dimensional scaffolds, providing mechanical support and guiding tissue formation (Vacanti and Langer, 1999). These scaffolds can be derived from natural or synthetic materials and are often seeded with cells or growth factors to enhance tissue regeneration and repair (Hutmacher, 2000). By harnessing the principles of tissue engineering, it is possible to develop advanced therapies, such as organoids, tissue constructs, and even whole organs, to address the growing demand for functional tissue replacements in regenerative medicine (O'Brien, 2011; Pina et al., 2019).

Designing engineered tissue for biomedical applications requires careful consideration of several vital factors. The choice of appropriate biomaterials and scaffolds is crucial; ideally, the biomaterials should be able to mimic the native extracellular matrix (ECM) to promote tissue regeneration. The ECM is a complex and dynamic network of macromolecules that surrounds and supports cells within tissues and organs, playing a vital role in tissue development, homeostasis, and cellular behavior (Figure 2). First and foremost, the ECM provides structural support and mechanical integrity to tissues, primarily attributed to collagens, the predominant ECM proteins (Frantz et al., 2010; Kular et al., 2014). Additionally, the ECM regulates cell behavior by providing biochemical and biomechanical cues through its interaction with cell surface receptors. It plays a vital role in cellular adhesion, migration, proliferation, and differentiation (Discher et al., 2009; Hynes, 2009).

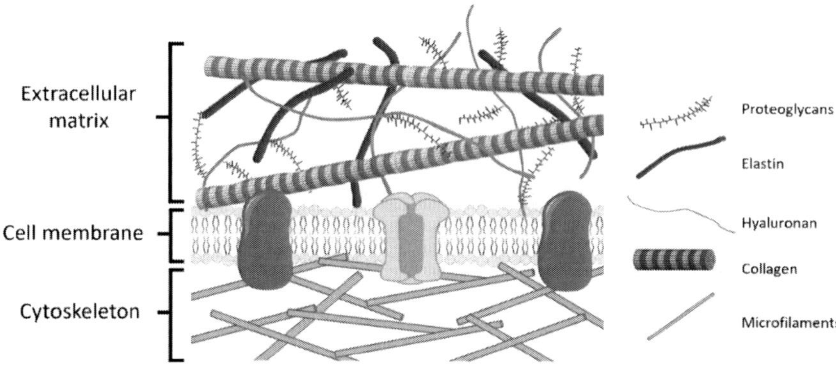

Figure 2. Graphical representation of extracellular matrix microenvironment and its main components. The ECM provides structural support and mechanical integrity to tissues, attributed mainly to the presence of collagens and polymeric substances that create a highly hydrated mesh.

Moreover, the ECM acts as a reservoir for growth factors, cytokines, and other signaling molecules, influencing their availability and activity (Lu et al., 2011). The ECM also participates in tissue remodeling and repair processes by regulating cell-mediated matrix degradation and synthesis (Karsdal et al., 2013). Furthermore, the ECM contributes to the modulation of immune responses and angiogenesis, essential processes in tissue homeostasis, and wound healing (Eming et al., 2007; Midwood et al., 2016). The ECM is a dynamic microenvironment that orchestrates cellular behavior and tissue function, making it a critical component in various physiological and pathological contexts. Gaining insights into the composition and structure of the extracellular matrix will facilitate the investigation of how it can be effectively employed as a scaffold in tissue engineering endeavors.

ECM comprises various constituents that interact to provide structural integrity, mechanical support, and biochemical signaling cues. Collagens are the most abundant ECM proteins that form the structural framework and impart tensile strength (Hynes, 2012; Ricard-Blum, 2011). Additionally, elastin, a highly elastic protein, confers resilience and elasticity to tissues (Mecham, 2018). Proteoglycans, such as aggrecan and decorin, are responsible for hydration and compressibility due to their ability to bind water molecules (Karamanos et al., 2019). Glycoproteins, including fibronectin and laminin, provide adhesive properties and cell attachment sites and regulate cellular activities (Mao and Schwarzbauer, 2005; Yurchenco et al., 2004). Together, these constituents form a complex and highly regulated network that governs tissue development, remodeling, and repair in various physiological and pathological contexts. Natural scaffolds have garnered significant attention in tissue engineering due to their inherent biocompatibility, bioactivity, and resemblance to the native ECM.

One of the main goals of tissue engineering is ECM mimicry, which provides a biomimetic environment that closely resembles the native ECM. ECM mimicry promotes cellular interactions by providing a familiar microenvironment that supports cell adhesion, migration, proliferation, and differentiation. The presence of ECM-like components in scaffolds enhances cellular attachment and facilitates the formation of functional tissue constructs (He et al., 2020). Also, ECM mimicry allows for the presentation of biochemical cues, such as growth factors, cytokines, signaling peptides, or small molecules, in a spatially and temporally controlled manner (Johnson et al., 2010; Taylor et al., 2006; J. Yang et al., 2023). These cues should be carefully considered since they can influence cell behavior, tissue maturation, and functionality (Gresham et al., 2021; He et al., 2020). These cues' delivery method, timing, and dosage should be optimized to achieve the desired tissue outcomes (Lee et al., 2019; Man et al., 2021). In addition, ECM mimicry can guide tissue-specific differentiation by providing tissue-specific cues, including biomechanical signals. Engineered constructs can generate tissue-specific architectures and functionalities by replicating the ECM characteristics of specific tissues, such as bone, cartilage, or muscle (Wang et al., 2023; K. Yang et al., 2023). The physical properties of the engineered tissue, such as scaffold architecture, porosity, and mechanical properties, should be designed to match the target tissue's specific requirements. Proper vascularization and nutrient supply are also critical, as they ensure cell viability and functionality within the engineered tissue (Collins et al., 2021; Guimarães et al., 2020). Therefore, ECM mimicry is considered a valuable tool in tissue engineering, providing a biomimetic environment that promotes cellular activities, tissue formation, and functional regeneration.

One widely used natural scaffold in tissue engineering and regenerative medicine is collagen, a major structural component of the ECM in various tissues (Parenteau-Bareil et al., 2010; Zihan Wang et al., 2021). Collagen is a fundamental protein crucial in the ECM of different animal tissues, including humans. It is the most abundant protein in the human body, comprising about 30% of the total protein content (Rezvani Ghomi et al., 2021b; Sherman et al., 2015). Collagen is known for its remarkable tensile strength and flexibility, imparting structural integrity and resilience to tissues such as skin, bones, tendons, ligaments, cartilage, and blood vessels. Structurally, collagen is a triple-helical protein composed of three polypeptide chains known as alpha chains, which are primarily rich in the amino acids glycine, proline, and hydroxyproline. These alpha chains form a right-handed helix, which further associates with other helices to create fibrils, the basic building blocks of the ECM (Ottani et al., 2001; Shoulders and Raines, 2009). The structural characteristics and biocompatibility of collagen make this natural biopolymer a good candidate for developing scaffolds for tissue engineering purposes. The structural features and biocompatibility of collagen make this natural biopolymer a good candidate for developing scaffolds for tissue engineering purposes.

Collagen-based scaffolds provide excellent biocompatibility, support cell adhesion, proliferation, and differentiation, and exhibit tunable mechanical properties (Copes et al., 2019; Parenteau-Bareil et al., 2010). These biomolecules have served as bioinspiration in the designing and manufacturing scaffolds for regenerative medicine and tissue engineering. For example, Cao et al. (2015) fabricated fish collagen/chitosan/chondroitin sulfate scaffolds incorporated with poly(lactide-co-glycolide) (PGL) microspheres loaded with essential fibroblast growth factor (bFGF). This growth factor plays a significant role in wound healing, particularly by stimulating angiogenesis and regulating multiple aspects of cellular functions, such as cell movement, growth, and the metabolism of the extracellular matrix (Cortina and Azar, 2010). The scaffolds exhibited an interconnected porous structure with the microspheres uniformly distributed that promotes the attachment and spreading of fibroblast cells on the surface and inside the pores of scaffolds. At the same time, swelling and degradation rate of scaffolds lead to diffusion of bFGF-loaded microspheres, promoting fibroblast proliferation. *In vivo* evaluation using a full-thickness skin wound model in rats showed that fibroblast cells increased after four weeks of collagen scaffold implantation, and the inflammatory reaction showed improvement. By the 6-week treatment, many fibroblasts had migrated into the implant sites, and the inflammatory cells were no longer detectable. These findings indicate that collagen-based scaffolds with PGL loaded with bFGFh benefit cell proliferation and tissue regeneration for skin tissue engineering (Cao et al., 2015).

Long et al. (2015) elaborated collagen-based membranes using silk fibroin for corneal tissue engineering. Silk fibroin was incorporated to improve the mechanical properties of collagen scaffolds to satisfy the requirements of clinical applications. The membranes' tensile stress and elastic modulus decreased as the silk fibroin content increased. Conversely, the elongation at the break of the membranes increased with the rise in silk fibroin concentration. *In vitro*, biocompatibility assays using human corneal epithelial cells (HCEC) showed that the cells adhere and spread on collagen membranes with silk fibroin (10%). The cells change their morphology to a spindle shape resembling normal HCECs. In addition, lamellar keratoplasty on rabbits showed that the transplants' transparency is rapidly regained within the initial 35 days after surgery, while the epithelial coverage was mainly accomplished. The implants exhibit excellent integration into the recipient's corneas, creating a smooth and seamless corneal surface. Also, corneal neovascularization or rejection reaction was observed within the 6-week transplantation period.

Another frequently employed biomaterial is hyaluronic acid (HA), known for its lubricating and hydrating properties. HA-based hydrogels have been used in tissue engineering applications, particularly cartilage regeneration (Humaira et al., 2022; Oliva et al., 2021). Additionally, gelatin, derived from collagen, has been widely used due to its structural similarity to collagen and ease of modification (Echave et al., 2017; Lukin et al., 2022). Natural polymers such as alginate, chitosan, and fibrin have also been extensively investigated for their biocompatibility and tissue-specific cues (Bujoli et al., 2019; Madni et al., 2021; Sahoo and Biswal, 2021). Furthermore, synthetic biomaterials like poly(lactic-co-glycolic acid) (PLGA) and poly(ethylene glycol) (PEG) offer tunable properties and controlled degradation rates, making them versatile options in tissue engineering (Chew et al., 2021; Li et al., 2023). These common biomaterials have significant potential in facilitating ECM mimicry and advancing tissue engineering approaches for diverse regenerative medicine applications.

Several techniques have been employed to fabricate scaffolds with diverse structures and properties suitable for tissue engineering applications. One commonly utilized technique is electrospinning, which involves using an electric field to generate ultrafine fibers from a polymer solution or melt. Electrospun scaffolds offer high surface area, tunable porosity, and controllable mechanical properties, making them ideal for various tissue engineering applications (Babitha et al., 2017; Iacob et al., 2020). Another prominent method is 3D bioprinting, enabling precise and controlled deposition of biomaterials and cells layer by layer to create complex 3D structures. This technique allows the incorporation of multiple cell types and biomaterials, enhancing tissue-specificity and functionality (Kang et al., 2016).

Additionally, freeze-drying or lyophilization is employed to create porous scaffolds from cryogels or hydrogels, offering interconnected porous networks that facilitate cell migration, nutrient diffusion, and tissue regeneration (Fereshteh, 2018; Nguyen et al., 2013). Furthermore, decellularization techniques have gained attention, where natural tissues or organs are treated to remove cellular components, leaving behind a scaffold with preserved ECM architecture. Decellularized scaffolds provide native ECM cues and have shown promise for tissue-specific regeneration (Garreta et al., 2017; Hoshiba et al., 2010). These techniques, among others, demonstrate the versatility and advancements in scaffold fabrication for tissue engineering, offering a wide array of approaches to meet the specific requirements of different tissues and organs.

Selecting techniques for scaffold fabrication in tissue engineering is a critical decision, and several vital criteria are considered to ensure the success and efficacy of tissue regeneration. A crucial criterion is the ability of the technique to create a scaffold with a suitable microarchitecture. Scaffolds should have an interconnected porous structure, allowing efficient nutrient and oxygen diffusion and supporting cellular infiltration and tissue integration (Ahn et al., 2010; Sachlos and Czernuszka, 2003). Additionally, the mechanical properties of the scaffold play a vital role in matching the mechanical demands of the target tissue. Scaffolds must have appropriate stiffness and elasticity to withstand physiological loads and provide mechanical cues that direct cell behavior. Moreover, the fabrication technique's scalability and reproducibility are essential for potential clinical translation and commercialization. Techniques that can produce scaffolds in large quantities with consistent quality are preferred. Furthermore, the capability to incorporate multiple bioactive molecules or growth factors is a significant advantage in promoting tissue-specific regeneration. Ultimately, the choice of the technique should align with the specific requirements of the target tissue, ensuring optimal scaffold design and functionality for successful tissue engineering outcomes.

Chemical Modification of Extracellular Polymer Scaffolds to Enhance Interactions with Cells

To adequately mimic the extracellular matrix, it is necessary to simulate its mechanical and structural properties and the physicochemical and biochemical signals required for the correct differentiation of cells during tissue development (Sgambato et al., 2016). The chemical properties of the biomaterials used in developing scaffolds are exploited to achieve this. These biopolymers can be chemically modified or functionalized by incorporating chemical or biological moieties onto the biopolymer's surface or structure to impart specific capabilities, such as controlled drug release, targeted delivery, improved cell adhesion, or enhanced mechanical strength (Figure 3) (Jing et al., 2023). This process can be achieved through various techniques, including covalent bonding, physical adsorption, or blending with other biomaterials or bioactive agents (Niemczyk-Soczynska et al., 2020). Functionalized biopolymers offer tailored properties that cater to the requirements of different biomedical applications, such as tissue engineering, drug delivery, wound healing, and medical device fabrication.

Modifying the mechanical properties of scaffolds used in tissue engineering is paramount to mimic the native tissue environment and achieve successful tissue regeneration. For instance, Jing et al. (2023) developed a nanofibrous scaffold topographical and chemical biomimicry to ECM. The scaffold employed a PCL nanofibrous substrate in a “shish-kebab” arrangement to achieve topographical similarity with the ECM. Then, it was coated with polydopamine and arginyl glycyl aspartic acid (RGD) to provide chemical biomimicry. The introduced shish-kebab structure and grafted RGD enhanced the surface roughness and improved the nanofibrous scaffold's surface wettability. In addition, the modified scaffolds promoted the growth of human umbilical vein endothelial cells. Also, Abbasnezhad et al. (2023) chemically modified an acellular fish skin scaffold via 1-ethyl-3-(3-dimethyl aminopropyl) carbodiimide (EDC) cross-linking to improve mechanical strength, denaturation temperature, and degradability. The maximum tensile stress of the cross-linked samples, with different cross-linker concentrations at 5, 10, 20, 30, and 40 mM, varying from 52.25 MPa (with Young's modulus of 809.26 MPa) to 175.31 MPa (Young's modulus of 1245.54 MPa). In contrast, the maximum tensile stress of the acellular fish skin sample without cross-linking was 47.94 MPa (Young's modulus of 739.06 MPa). In addition, scaffold modification with EDC cross-linker increases thermal stability from

58 °C for uncross linked samples to 87 °C for the cross-linked sample. Also, samples modified by higher cross-linker concentrations (20 - 40 mM) presented less weight loss and maintained stability for up to 14 days. Therefore, the authors concluded that employing the carbodiimide cross-linker can enhance the properties and performance of biological tissues, such as acellular fish skin.

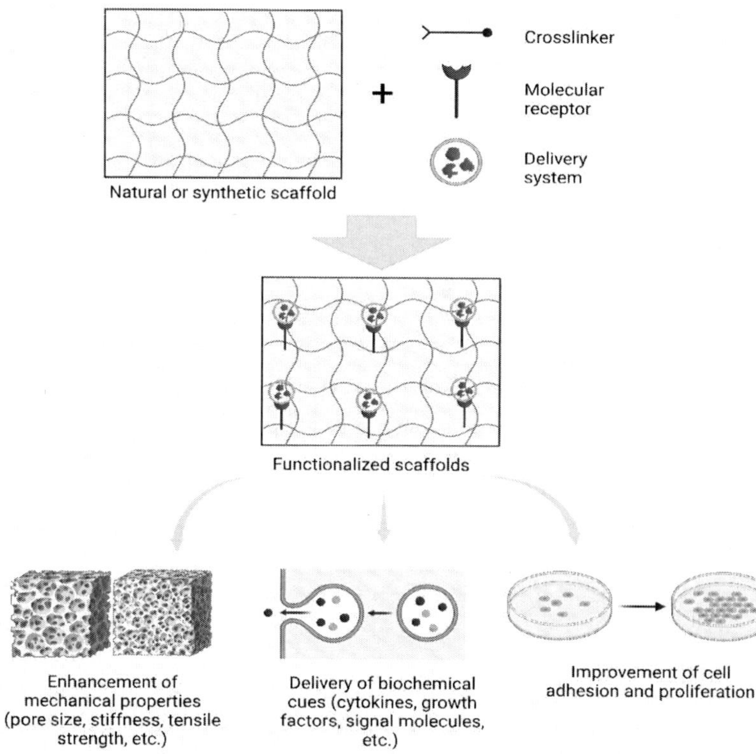

Figure 3. Schematic representation of chemical modification of extracellular polymer scaffolds to improve mechanical properties and to incorporate biochemical cues delivery systems to enhance cell adhesion and proliferation.

Several studies have highlighted the significance of scaffold mechanical properties in influencing cell behavior, tissue formation, and overall tissue-engineered construct performance. Adequate mechanical properties are essential to mimic the native tissue environment and achieve successful tissue regeneration. For instance, studies have shown that the stiffness of the scaffold can influence stem cell differentiation, with softer scaffolds promoting neurogenic differentiation and stiffer scaffolds favoring osteogenic or chondrogenic differentiation (Discher et al., 2009; Guilak et al., 2009; Huang et al., 2015). Scaffolds' alignment and topographical features have also been demonstrated to affect cell morphology, migration, and tissue organization (Bettinger et al., 2009). Additionally, scaffold porosity and pore size can impact nutrient diffusion, oxygen availability, and cell infiltration, ultimately influencing tissue ingrowth and vascularization (Li et al., 2016). Overall, optimizing the mechanical properties of scaffolds is essential for guiding cellular responses and promoting successful tissue development in tissue engineering applications.

Another target of chemical modification of extracellular polymers is to enhance the interactions of scaffolds with cells. For example, Aldemir Dikici et al. (2020) functionalized a four-arm hydroxyl-terminated polycaprolactone (4PCL) with methacrylate to make the polymer photocurable. With this material, the authors elaborated multiscale porous polycaprolactone (PCL)-based high internal phase emulsions (HIPE) as a bone repair material. The material was seeded with bone cells to decorate these scaffolds with an *in vitro* cell-derived ECM. After the decellularization of these constructs, a biohybrid scaffold was obtained. Cell bone-derived ECM deposition onto the polymeric scaffold increased the initial cell attachment up to 3-fold compared to the PCL-only scaffold. In addition, biohybrid scaffolds showed better cell attachment, proliferation, and ECM deposition ability of human embryonic stem cell-derived mesenchymal progenitor cells.

Similarly, Chernozem et al. (2020) explored the chemical modification of piezoelectric polyhydroxy butyrate (PHB) and non-piezoelectric PCL scaffolds to improve osteoblast adhesion and proliferation. By modifying the scaffold surfaces with 3,4-dicarboxybenzenediazonium tosylate (ADT-(COOH)2) under ultraviolet irradiation, they achieved improved wettability and osteoblasts adhesion on both piezoelectric PHB and non-piezoelectric PCL scaffolds. The modified scaffolds showed a significant decrease in water contact angle, transforming them from hydrophobic to hydrophilic without compromising their intrinsic structure. The scaffolds retained their piezoelectric response, and osteoblastic cells exhibited enhanced spreading and proliferation on the treated surfaces compared to pristine scaffolds. The findings suggest that this facile surface modification technique using ADT-(COOH)2 holds promise for enhancing the properties and performance of biodegradable piezoelectric scaffolds in tissue engineering applications.

In another study, Idini et al. (2019) developed PCL scaffolds functionalized with glycosaminoglycans (GAGs) to support Schwann cells, a glial cell type in the peripheral nervous system that play essential roles in the development, maintenance, function, and regeneration of peripheral nerves (Lavdas and Matsas, 2014). Schwann cells were seeded on functionalized and nonfunctionalized scaffolds with random or aligned fibers. After 1, 3, and 7 days of culture, metabolic activity, proliferation, and GAGs assays were performed. Furthermore, the expression of specific markers was assessed by immunofluorescence and western blot analyses. Among them, syndecan 1, a transmembrane proteoglycan involved in cytoskeletal reorganization and the attachment of cells to ECM, and syndecan 4, which binds different proteins in ECM and promotes focal adhesion in co-localization with integrins.

Additionally, the authors evaluated the expression of p75, a marker of differentiation in Schwann cells, and laminin, one of the most abundant ECM proteins in the peripheral nervous system. The results showed that GAG-functionalized scaffolds increased Schwann cells' metabolic and proliferation activities over seven days in culture and GAG production. Also, GAG-functionalized aligned fibers resulted in higher expression of syndecan four than GAG-functionalized random fibers and nonfunctionalized scaffolds, indicating that fiber topography and GAG functionalization modulated syndecan four productions. These observations suggested that GAG-functionalized scaffolds may create an extracellular environment similar to the natural Schwann cell basal lamina that could help to improve peripheral nervous system repair (Idini et al., 2019).

Extracellular polymer scaffolds can also be versatile carriers for delivering biochemical cues, including growth factors, signaling peptides, cytokines, or small molecules that can influence cell behavior, tissue maturation, and functionality. Ravindran Girija et al. (2020) functionalized electrospun PCL scaffolds via aminolysis surface modification and NaOH treatments to attach ECM proteins, collagen, and fibronectin. The functionalized PCL scaffolds were used to connect osteogenic extracellular vesicles (EV) collected from mechanically stimulated MLO-Y4 osteocytes. EVs are highly effective intercellular communication vehicles, as they can transfer RNAs, proteins, and lipids, playing crucial roles in various biological processes across different biological systems (Doyle and Wang, 2019). EV functionalization of PCL electrospun scaffolds enhanced early osteogenesis and total mineral content. The data suggest that collagen and EV functionalization have the potential to promote osteogenesis and improve mineralized matrix deposition in tissue engineering applications (Ravindran Girija et al., 2020).

Hao et al. (2020) immobilized mesenchymal stem cells derived EVs (MSC-EVs) onto electrospun PLLA/PCL nanofibrous scaffold by using LLP2A, an integrin α4β1 ligand, which had a strong binding to human placenta-derived MSCs, as the conjugation linker to construct a functional biomaterial scaffold. The results showed that MSC-EVs functionalized PLLA/PCL electrospun scaffolds increased the expression of human umbilical vein endothelial cells' angiogenic genes, KDR and TIE2 compared to the untreated PLLA/PCL electrospun scaffolds. Also, MSC-EVs functionalized PLLA/PCL electrospun scaffolds improved human umbilical vein endothelial cells' survival, decreased the apoptotic rate, and downregulated the expression of apoptotic markers (caspase nine and caspase 3). These findings suggest electrospun scaffolds modified with PMSC-EVs are promising for vascular tissue regeneration (Hao et al., 2020).

In the same way, K. Yang et al. (2023) fabricated an aligned fibrin nanofiber hydrogel modified with N-Cadherin-Fc (AFGN) by electrospinning and biochemical conjugation to deliver neural stem cells for spinal cord injury repair. Aligned fibrin nanofiber hydrogel (AFG) was generated by electrospinning, and their COOH groups were activated and stabilized in an ethyl-3-(dimethyl aminopropyl) carbodiimide hydrochlorid/N-hydroxysuccinimide solution (0.4/0.6 mg/mL) and coated with 2 μg/mL protein A solution. Finally, the nanofibers were functionalized with recombinant human N-cadherin human Fc chimera protein. Rheological characterization showed that unmodified and modified aligned fibrin nanofiber hydrogels exhibited viscoelasticity properties that resemble soft tissues. Also, hydrogels exhibited Young's moduli of 1.22 ± 0.17 kPa and 1.18 ± 0.06 kPa, respectively, which were in the range of the moduli of the ECM in native nerve tissue (~1–4 kPa). In addition, the functionalization with N-Cadherin-Fc enhanced the hydrophilicity of the aligned fibrin nanofiber hydrogels. The designed AFGN hydrogel exerted a combination of cues, including aligned nanofibrous topography, soft mechanical properties, and specific cell binding ligands, to guide neural stem cell behavior and promote nerve regeneration effectively. Incorporating N-Cadherin-Fc enhances cell adhesion and mechano-sensing of neural stem cells on the AFGN substrate, leading to favorable differentiation outcomes. In a rat model of complete transected spinal cord injury, the AFGN hydrogel and exogenous neural stem cells demonstrated remarkable effects, including improved cell retention, immunomodulation, neuronal differentiation, and integration with existing neurons. This approach facilitated the formation of a regenerated neural relay and showed promising results in locomotor functional recovery.

Wang et al. (2023) fabricated small-diameter vascular grafts with biomimetic intimal niche architecture. These biomaterials were made by electrospinning using polylactic acid and silk fibroin solutions. These vascular drafts were functionalized with copper ions (Cu^{2+}) chelated dopamine to catalytic release nitric oxide signaling molecules as a biochemical cue for regulating multiple vascular cell behaviors of endothelial cells, smooth muscle cells, and platelets. Incorporating dopamine or dopamine-chelated copper ions (Cu^{2+}) into vascular draft does not modify the biomaterials' mechanical properties (tensile strength, elongation at break, burst pressure suture retention, and compliance).

On the other hand, the assessment of vascular cells, such as endothelial cells and smooth muscle cells, demonstrates that vascular drafts functionalized with copper ions (Cu^{2+}) chelated dopamine can efficiently suppress the expression of von Willebrand Factor (vWF), leading to increased proliferation of endothelial cells, and simultaneously enhance the expression of α-SMA, inhibiting the proliferation of smooth muscle cells. Both effects are advantageous in regulating vascular homeostasis. The results of *in vivo* assessment using a rat abdominal aorta model showed functionalized vascular grafts promoted tissue formation with an intact and dense neointima, resembling the structure of native vessels. In contrast, the control groups are entirely blocked by severe thrombus. Histological observations also reveal positive staining for ECM components, including elastin fibers, muscle fibers, and collagen, in the neointima of vascular drafts functionalized with copper ions (Cu^{2+}) chelated dopamine. The tube wall of these grafts is slightly thicker than native vessels due to the longer degradable time of the primary material. These results indicate that biological vascular grafts made of biocompatible materials can effectively remodel the regeneration of damaged vessels (Wang et al., 2023).

The studies discussed in this section demonstrate the importance of tailoring scaffolds' physicochemical and biochemical properties to mimic the native extracellular matrix and create a conducive environment for cell differentiation and tissue development. The success of this strategy provides valuable insights for designing and developing advanced biomaterials that hold great potential in tissue engineering, wound healing, and medical device fabrication. However, further research is still needed to fully understand these modified scaffolds' long-term effects and translational applications in tissue engineering and regenerative medicine.

Incorporation of Peptic Sequences into Extracellular Polymers to Improve Bioactive Properties and Scaffold Functionality

Extracellular polymers (EPS) have garnered considerable attention in tissue engineering and regenerative medicine due to their inherent biocompatibility and ability to mimic the natural extracellular matrix (Osmani et al., 2021). However, researchers have incorporated peptic sequences, short bioactive peptides, to maximize their potential for biomedical applications into EPS scaffolds (Tan et al., 2020). The strategic integration of these bioactive motifs within the polymer matrix offers a promising approach to improve the bioactive properties and scaffold functionality, thereby enabling better cellular interactions and tissue regeneration. Therapeutic peptides encompass diverse molecules that play crucial roles as hormones, growth factors, neurotransmitters, ion channel ligands, and anti-infective agents (Wang et al., 2022). These peptides exert their effects by binding to specific cell surface receptors, triggering well-defined intracellular signaling pathways with high affinity and remarkable specificity (Muttenthaler et al., 2021). Therapeutic peptides operate similarly to biologics, including therapeutic proteins and antibodies, as they interact with their molecular targets to achieve therapeutic outcomes.

One advantage of therapeutic peptides over biologics is their reduced immunogenicity, making them less likely to elicit adverse immune responses in patients (Bashor et al., 2022). This characteristic is particularly advantageous for long-term therapeutic applications, where maintaining treatment efficacy without significant immune-related complications is crucial. Moreover, therapeutic peptides often boast lower production costs than larger biologics, a benefit that translates into more accessible and affordable treatments (Tombling et al., 2021). This cost-effectiveness is significant for healthcare systems, potentially broadening patient access to novel peptide-based therapies. For this reason, peptic sequences have emerged as crucial regulators of cell behavior, exerting their influence through interactions with specific cell receptors and extracellular matrix proteins (Rodrigues et al., 2007).

Table 1 shows reported peptic sequences with the potential to improve scaffold functionality. For example, the Arg-Gly-Asp (RGD) motif is pivotal among the extensively studied peptic sequences. This motif selectively binds to integrin receptors on the cell surface, leading to a cascade of events promoting cell adhesion, migration, and tissue formation (Xian et al., 2010). By incorporating RGD or other cell-adhesive sequences into EPS scaffolds, tissue engineers create a highly favorable microenvironment, fostering the anchorage and proliferation of specific cell types. This orchestrated cellular response significantly enhances the formation of functional tissue constructs, holding immense potential for diverse biomedical applications.

Table 1. Examples of peptic sequences have been used to improve the bioactivity of EPS scaffolds for various biomedical applications

Peptic sequence	Scaffold bioactive properties	References
Arg-Gly-Asp (RGD)	Incorporating RGD peptide into EPS scaffolds enhances cell adhesion, migration, and tissue formation, making it valuable for tissue engineering applications, such as bone and cartilage regeneration. RGD promotes the recruitment and adhesion of keratinocytes and fibroblasts, accelerating the formation of granulation tissue. These benefits highlight the potential of EPS scaffolds with RGD peptides for regenerative medicine and wound healing.	(Li et al., 2022)
Ile-Lys-Val-Ala-Val (IKVAV)	Incorporating the IKVAV peptide into natural or polymeric scaffolds has enhanced stem cell interactions, cell adhesion, neuron differentiation, and neurite extension, essential for neuron tissue regeneration. This combinatorial approach allows tailored bioactive scaffolds suitable for nerve tissue engineering. IKVAV improves cell-ECM interaction and promotes neuronal growth when applied *in vivo*, reducing glial cells. Its versatility as a nanofiber and a hydrogel makes IKVAV a promising biomaterial for stem cell therapy and neuron regeneration, offering exciting possibilities for advancing neuroregenerative strategies.	(Patel et al., 2019)
Tyr-Ile-Gly-Ser-Arg (YIGSR)	YIGSR has demonstrated significant potential in promoting essential cell behaviors such as attachment, proliferation, and migration. By incorporating YIGSR into EPS scaffolds, researchers have observed notable improvements in endothelial cell adhesion and angiogenesis. This makes it a promising strategy for applications in vascular tissue engineering and wound healing.	(Tsur-Gang et al., 2009)

Peptic sequence	Scaffold bioactive properties	References
	YIGSR has been successfully employed to modify polysaccharide hydrogels, effectively enhancing a wide range of cell behaviors, including cell survival, adhesion, proliferation, and migration. This versatile peptide-based modification holds great promise for advancing biomaterials in regenerative medicine and tissue engineering, offering exciting prospects for developing innovative therapies and interventions to promote tissue repair and regeneration.	
Leu-Asp-Val (LDV)	LDV sequence has proven to be a suitable peptide for promoting cell migration and tissue repair. EPS scaffolds enhanced with LDV have remarkably improved cell infiltration and tissue integration, making them valuable assets in wound healing and tissue regeneration studies. LDV peptide coating has exhibited a dual benefit, enhancing endothelial cell adhesion on ePTFE disks *in vitro* and contributing to the regeneration of neointima-like tissue on the intimal surface of ePTFE patches *in vivo*. These findings highlight the potential of LDV as a bioactive coating for ePTFE surfaces, offering exciting prospects for promoting tissue repair and regeneration.	(Kakinoki et al., 2022)
Gly-Arg-Gly-Asp-Ser-Pro (GRGDSP)	GRGDSP sequence, derived from fibronectin, has been well-documented for its high affinity toward integrin receptors present on the surfaces of various cell types, including endothelial cells (ECs), fibroblasts, and smooth muscle cells. This peptide holds immense potential for functionalizing the external side of scaffolds, promoting robust adhesion of fibroblasts and smooth muscle cells. By creating a bi-functionalized scaffold tailored to engage specific cell types, the regeneration of a layered tissue that closely resembles the native structure of small-caliber arteries becomes feasible.	(Rosellini et al., 2022)

Nowadays, the incorporation of peptic sequences into EPS scaffolds represents a powerful strategy for enhancing their bioactive properties and scaffold functionality. Recent research findings and advancements in this field underscore the potential of peptic sequence-modified EPS scaffolds for tissue engineering, regenerative medicine, and personalized healthcare. By leveraging the unique capabilities of these short bioactive peptides, tissue engineers can create tailored microenvironments that promote cell interactions, support tissue regeneration, and ultimately contribute to the development of innovative therapeutic approaches for various medical conditions. As research progresses, peptic sequence-incorporated EPS scaffolds are expected to significantly shape the future of biomedical applications revolutionizing regenerative medicine and personalized healthcare.

Controlled Drug Release from Extracellular Polymer Scaffolds for Therapeutic Applications

Controlled drug release from EPS has garnered significant attention in therapeutics due to its potential to revolutionize drug delivery strategies. These scaffolds offer unique advantages, such as high biocompatibility, mechanical properties, and the ability to provide a three-dimensional microenvironment for cells (Calori et al., 2020). Incorporating drug molecules into these scaffolds makes achieving sustained and localized drug release possible, optimizing therapeutic outcomes while minimizing potential side effects (Kabu et al., 2015). This section will explore strategies for controlling drug release from EPS scaffolds, including physical and chemical approaches.

Encapsulation is a fundamental approach to achieving controlled drug release from EPS, involving the entrapment of drug molecules within the scaffold matrix through physical mechanisms like diffusion or adsorption (Orive et al., 2014). Various encapsulation methods have been explored to control drug release kinetics precisely, tailoring drug-loaded EPS to meet specific therapeutic requirements. For instance, in a study by Cheng et al. (2020), the anti-cancer drug paclitaxel was encapsulated within an extracellular polymer scaffold using an emulsion technique, resulting in sustained release of the drug over an extended period, effectively inhibiting tumor growth. In another research by Qin et al. (2019), coacervation was employed to encapsulate insulin in an extracellular polymer scaffold, enabling the controlled release of the hormone to manage blood glucose levels in diabetic murine models. In an Elhassan et al. (2022) study, spray drying was utilized to encapsulate antibiotics in an extracellular polymer scaffold, achieving a prolonged and controlled release of the drugs and effectively combating bacterial infections.

Also, hydrogel formation with EPS presents another path for controlled drug release, where hydrogels absorb and retain significant amounts of water, creating an optimal environment for drug dissolution and diffusion (Mushtaq et al., 2022). For example, in a study by Grimaudo et al. (2022), extracellular polymer hydrogels loaded with nerve growth factors were engineered with stimuli-responsive components, responding to local pH changes in the injured neural tissue. The hydrogels demonstrated on-demand drug release, promoting nerve regeneration and functional recovery after injury. Similarly, in a study by Vismara et al. (2020), extracellular polymer hydrogels incorporated with nano-composite gel combining PEG, polyethylene-imine (PEI) conjugated with Rolipram were developed, allowing the release of anti-inflammatory drugs to target the astrocytes inflammatory response to injuries. By harnessing the capabilities of controlled drug delivery molecules in extracellular polymer scaffold hydrogels, researchers can design innovative drug delivery systems with tailored release profiles, enhancing therapeutic efficacy and improving patient outcomes across various biomedical applications.

On the other hand, chemical strategies play a crucial role in achieving controlled drug release from extracellular polymer scaffolds, offering precise and suitable drug therapeutic applications using covalent conjugation by electrostatic interactions (Calori et al., 2020);. In this context, the covalent bonding of drug molecules to EPS chains provides a robust and stable drug delivery mechanism, offering precise control over drug release rates employing linker molecules with tunable degradation rates; researchers can design drug-polymer conjugates that release drugs through hydrolysis or enzymatic cleavage at specific rates (Muttenthaler et al., 2021). This strategy has several advantages, including precise control of drug quantity, strong stability of the drug within the matrix, prevention of drug aggregation, quenching of burst release, and avoidance of drug loss during scaffold sterilization or purification. Additionally, drug-polymer conjugation efficiently decreases drug release rates compared to physical loading approaches, making it an innovative and highly efficient strategy for increasing drug loading (Park et al., 2019). For example, Nosrati et al. (2019) evaluated biotin-PEG-PCL copolymers that demonstrated significant potential in targeted drug delivery to cancer, enhancing drug pharmacokinetics and mitigating adverse effects. For this reason, incorporating these chemical strategies into extracellular polymer scaffolds opens exciting possibilities for advanced drug delivery systems with precisely controlled release profiles. Continued research and development in physical and chemical strategies for controlled drug release from extracellular polymer scaffolds will undoubtedly advance the field of drug delivery, paving the way for innovative and more effective therapeutic interventions.

Use of Extracellular Polymers in Disease Modeling and Evaluation of New Therapies

Extracellular polymeric substances are components used in various treatments (wastewater, medical, and in the preparation of new bioengineered materials) and applications due to their physicochemical and biological properties that include polysaccharides, proteins, lipids, and extracellular DNA (Liu et al., 2018). This section will focus on the promising applications it has in medical areas; one of the most relevant EPS is polyacrylic acid, which is non-toxic, biocompatible, and biodegradable; it is also marketed at a low price and approved by the FDA (Savaskan Yilmaz et al., 2020). Yan et al. (2022) developed a polyacrylic acid hydrogel incorporated in polydopamine/gelatin complex (PDA/gel-PAA) and evaluated its effectiveness in the regeneration of cartilage in 6 rabbits; this hydrogel presented thermally more stable, resistant, and adhesive characteristics for tissues, this may be because it has a more significant amount of carboxyl, amino and catechol groups, as well as the multiple non-covalent interactions in the PDA/gel complex and the PAA network contributed to building a hybrid physical-chemical crosslinking network with more excellent thermal stability in the gelatin chains which can be integrated with the tissue after implantation (Han et al., 2017). Likewise, the authors mention that it facilitated the proliferation of stromal cells in the bone marrow and promoted cartilage repair in encapsulating growth factors.

In the medical area, the combination of different EPS has been used to enhance its effects; Ni et al. (2023) developed a polyethyleneimine/polyacrylic acid/chitosan hydrogel (PEI/PAA/CS) to reduce ulcerative arterial bleeding which has high mortality and has been a challenge for medicine, they observed that in rats and rabbits. This hydrogel has rapid gelation, tissue adhesion, gastric acid resistance, hemostasis, and tissue repair, attributed to adding chitosan because it builds a system of electrostatic interaction and hydrogen bridges. What promotes wound repair and the ability to stop bleeding (Peng et al., 2021). Therefore, developing hydrogels with EPS presents an opportunity to innovate the design of these with improved biological functions and wound regeneration properties. Likewise, chitosan and hyaluronic acid are widely used in medicine; they are natural polymers with anti-inflammatory, healing, and tissue-regenerating properties, repairing the skin (Lou et al., 2019).

Soriano-Ruiz et al. (2020) designed a hydrogel of poloxamer (18%), chitosan (0.5%), and hyaluronic (0.2%), which included vitamins A, D, and E, for the treatment of skin burns, in this study 10 volunteers were recruited to apply 0.5 mL/cm^2 of the hydrogel in the participants, where no signs of redness or skin reactions were shown. They also used an induced model of skin burn in mice to which samples of the hydrogel were applied; histological studies showed a regenerative action on burns; these properties are attributed to hyaluronic acid and chitosan, the latter the authors proposed that it is a possible strategy to improve adhesiveness and residence time of poloxamer systems. The results were favorable regarding the microbial efficacy of this hydrogel because it has been reported that chitosan can exert these properties by destabilizing microbial membranes and a synergistic effect by adding vitamins. The authors propose that this hydrogel be used for research in later clinical stages due to the favorable results shown in this study.

Another medical application in which they have been used is bone regeneration; hyaluronic acid is cytocompatible and biodegradable. These characteristics make it widely used in preparing hydrogels and administering drugs, and carboxymethyl chitosan has a solubility in water and excellent biocompatibility (Chen et al., 2012). Tan et al. (2022) developed an injectable cell hydrogel based on carboxymethyl chitosan and hyaluronic acid containing hydroxyapatite; this hydrogel achieved high hydrophobicity and water retention capacity and also showed self-repair properties due to the formation of dynamic imine bonds. The cytocompatibility was evaluated with mouse fibroblast cells in concentrations of 25, 50, and 100 mg/mL, where it was observed that there was no visible toxic reaction in the cells, and the fibroblasts maintained a normal morphology with the coculture of the hydrogel extract. These results may contribute to the development of scaffolds for bone tissue; however, these results need to be tested in an *in vivo* model.

Likewise, Krishnan et al. (2021) developed a topical treatment based on the combination of drugs doxorubicin and camptothecin and a conjugate of hyaluronic acid where a synergistic potency can be observed; this treatment presented minimum toxicity for primary human epidermal keratinocytes *in vitro*, it has been reported that hyaluronic acid binds specifically to CD44. This cell surface receptor has an elevated expression in different types of cancers, including skin (Platt and Szoka, 2008). As for its effect *in vivo* studies, it was possible to confirm its efficacy in treating lesions without irritation or systemic absorption. I remove cancer with a selectivity of up to 21 times on healthy skin tissue. One of the side effects patients have when receiving radiotherapy are radiodermatitis which can cause swelling risk of infection. Presta et al. (2019) evaluated the efficacy of a hyaluronic acid gel for 20 days for the treatment of radiodermatitis in a clinical case of a patient who was treated with head and neck radiotherapy; the authors reported that the patient presented a drastic reduction in skin toxicity, as well as a decrease in pain, these results were obtained using a Radiation Toxicity/Oncology Grading scale as well as a semi-structured interview. Therefore, the use of hyaluronic acid may be a new therapy for the treatment of cancerous lesions.

Collagen has been used primarily as a drug transporter, tissue engineering material, bone filler, and burn and wound dressing (Wang et al., 2019). There are several sources of collagen; however, diseases in cows and pigs and religious beliefs have limited the mammalian derivative. Furtado et al. (2022) developed an extraction to obtain fish collagen due to its ease, high content, excellent absorption properties, low molecular weight, and biocompatibility. The authors report that one of the applications may be in corneal tissue regeneration because freeze-dried marine collagen scaffolds formed from fish scales have better physical, mechanical, and microbial properties. Without donor corneas, corneal implants based on marine collagen are biocompatible, capable of transmitting light, and have scattering values (Zhao et al., 2019). Therefore, this collagen could be an option with unique properties, a lower risk of disease transmission to humans, and easy absorption. However, this requires further *in vivo* studies.

Collagen has also been used for skin regeneration; Wang et al. (2020) developed a collagen matrix to accelerate wound healing produced from pig skin. The collagen produced showed that the chemical content was collagen type 1, and the characteristics were similar to those of humans; it was also determined that it was not toxic and exhibited biocompatibility in both *in vitro* and *in vivo* studies; this may be due to porosity which plays an essential role in the effectiveness of cell seeding, diffusion and also in the mechanical residence since this allows the cells to be evenly distributed and maintain a massive transpose of nutrients, oxygen, growth factors and also waste products. The scaffold possesses the stiffness to withstand the surrounding mechanical stresses during tissue neogenesis, which can provide a physical and biochemical environment similar to native tissue, which promotes cell adhesion, proliferation, and tissue neogenesis (Qiu et al., 2013). Therefore, this collagen matrix could be suitable for the reconstitution and regeneration of different wounds.

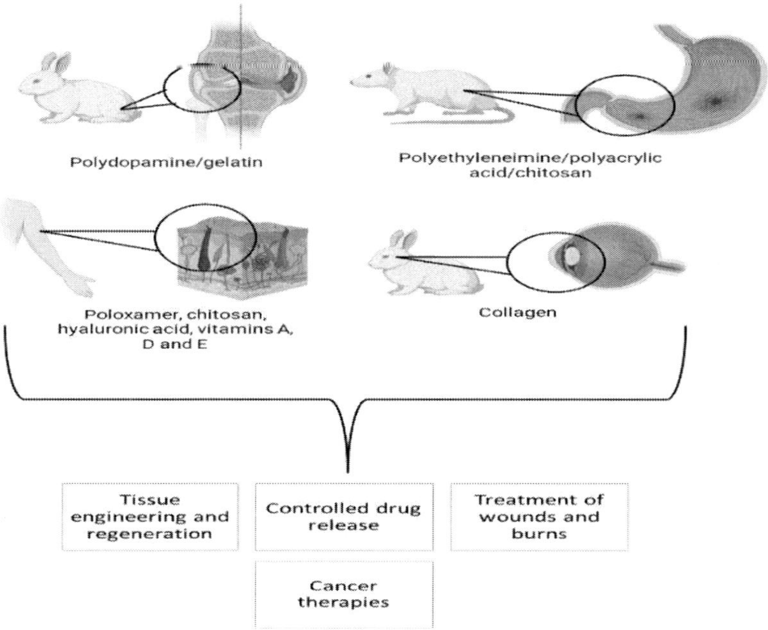

Figure 4. Applications of extracellular polymers in medicine.

Polylactic acid sutures are widely used in medicine because the human body can absorb them, are harmless, biodegradable, and have high resistance. Liu et al. (2019) determined the degradation period of sutures with polylactic acid and carbon nanotubes. They tracked and analyzed the structure and properties of the suture during its degradation in an *in vitro* model. The results demonstrated a delay in degradation and prolonged the force validity time in suture degradation. Also, Raynald et al. (2019) analyzed the efficacy of a nanofibrous polypyrrole/polylactic acid scaffold grafted into a spinal cord to observe if it promotes functional recovery in a spinal cord injury in rats; the results found by these authors are that these scaffolds have great potential to restore electrical conduction and promote functional recovery by inhibiting the formation of healing, promote axon regeneration and myelination in the area of injury. They also observed new vascular formation after implantation, decreasing neural apoptosis. This may be because the scaffold fills the lesion defect and reduces scar formation, which benefits axonal regeneration and remyelination; this effect has been previously reported in polylactic which can promote nerve regeneration, nerve conduction velocities (Mihardja et al., 2008).

Therefore, using it is essential to recover the electrophysiological properties of the nerve and achieve nerve conduction velocities within the range of the healthy nerve and the polylactic is a polymer polycationic duct that can attract cells to the area of injury (Wenjin et al., 2011). Electrical conduction carried by the scaffold can activate voltage-gated ion channels in cell membranes and induce modulation in intracellular signaling pathways, which can affect cell behavior, such as cell apoptosis (Widenfalk et al., 2003). Therefore, using these scaffolds to recover a spinal cord injury may be an alternative therapy. However, it is required to perform these studies in humans and measure whether the effect was similar to those in this study. Polyglycolic acid is widely used in the medical area, Fujimaki et al. (2019) designed an artificial nerve duct hybrid polyglycolic acid with dedifferentiated cellular fats, which they grafted into a murine model, observed that these promote nerve regeneration in an experimental model of rat nerve defects. The use of extracellular polymers in new therapies and applications in the medical area is becoming increasingly important; even more *in vivo* studies are required, but these advances could have applications in the future (Figure 4).

Combining Extracellular Polymers with Bioprinting Techniques for the Manufacture of Three-Dimensional Tissues

Today, three-dimensional (3D) bioprinting stands out as a field of research in tissue engineering, providing opportunities for personalized attention and a better understanding of cell-material interaction. Three-dimensional bioprinting is an innovative technique that enables the precise and controlled fabrication of tissue structures. The technology uses design software and cutting programs to divide and print cross-sections of a system sequentially, the creation of biological-synthetic hybrid scaffolds with suitable mechanical properties and specific cell organization. 3D printing stands out for its ability to design complex structures and the possibility of printing cells and biomaterials simultaneously. Although advances have been made in 3D bioprinting, building fully functional tissues/organs remains challenging (O'Brien et al., 2015). The technology faces limitations in mimicking natural structures and integrating them with native tissues. The utilization of advanced biomaterials and the focus on fabricating microstructures that mimic native tissues are active research areas (Luo et al., 2018). 3D bioprinting has the potential to revolutionize tissue engineering and regenerative medicine, and biomaterials play a critical role in its development and clinical application

Table 2. Examples of 3D printing for various tissue and organ regeneration

Tissue types	Printing materials	Description	References
Cartilage and osteochondral	Alginate hydrogel	Two layers of an osteochondral scaffold were fabricated and evaluated both *in vitro* and *in vivo*	(Fedorovich et al., 2012)
Neural	Hyaluronic acid hydrogel	Biomimetic nerve conduits have been created that have been shown to support neuronal development and the expansion of axonal processes in laboratory settings.	(Suri et al., 2011)
Complex tissue and organ	Alginate spheroids	Stem cells were imprinted similar to an organ mimetic. Cell viability was high and the spheroids fused together over time, adopting a continuous geometry.	(Williams et al., 2013)
Complex tissue and organ	Calcium chloride/ sodium alginate hydrogel	Three types of cells were simultaneously printed on a single support structure. All cell types maintained their viability and proliferative capacity, as well as their phenotypic expression and physiological function. In addition, vascularization was observed *in vivo*.	(Xu et al., 2013)
Retinal tissue	Alginate hydrogels	Desarrollo de human pluripotent stem cells in retinal tissue development	(Hunt et al., 2017)

Within 3D bioprinting technologies, different biomaterials, including extracellular polymeric substances, have been used to manufacture 3D tissue constructions. Conventional 3D bioprinting approaches fall into three main groups: inkjet, microextrusion, and laser-assisted bioprinting. Integrated extrusion-based 3D bioprinting has demonstrated the ability to print vascular networks within a support matrix. In most cases, the proper material selection is the most essential factor for each approach (Malda et al., 2013). The printability depends on the physicochemical properties of the pre-gel solutions and the gelling process. In recent decades, several extracellular polymeric substances have been used in 3D bioprinting (Table 2). However, the most used to obtain scaffolds and hydrogels are alginate, hyaluronic acid, and gellan (Bashiri et al., 2020). Alginate, in particular, is synthesized by certain bacterial species (*Pseudomonas aeruginosa* and *Azotobacter vinelandii*) and has been used in tissue engineering due to its biocompatibility, low toxicity, and ability to rapidly gelle through the addition of divalent cations such as Ca^{2+} (Sayyar et al., 2015). Alginate is also used with other hydrogenation materials, such as hydroxyapatite and gelatin (Akbardoost et al., 2018).

Creating three-dimensional structures of biologically relevant tissues is critical for alternative organ transplantation approaches in regenerative medicine and improving drug discovery trials (Yanagawa et al., 2016). In recent years, hydrogels have been used to reconstruct *in vitro* models in tissue engineering to mimic native tissue. These hydrogels can encapsulate bioactive molecules and cells and efficiently transfer mass by diffusion and biocompatibility (Peppas et al., 2006). Hydrogels are natural materials, such as alginate, gelatin, and hyaluronic acid, which help investigate cell-cell and cell-extracellular matrix interactions (Li et al., 2012). Extracellular polymers have shown promise in tissue/organ manufacturing and improving print resolution. Despite achievements in reproducing tissue and organ models, such as the heart, the full functionality of tissues remains challenging due to their complexity and composition (Jeon et al., 2019). Skeletal muscle tissue is composed of long multinucleated fibers arranged in parallel. Instead of using the individual's muscle tissue to treat muscle volume loss, skeletal muscle tissue engineering is needed to mimic the function and properties of natural muscle tissue. 3D bioprinting technologies are an ideal choice, as they enable the precise fabrication of similar biomimetic structures. In the 3D bioprinting of skeletal muscle tissue, different bioinks present high cell viability, differentiation, maturation, and contractibility, making them promising biomaterials for this type of tissue (Davoodi et al., 2020). In cell alignment studies for developing biomimetic muscle constructs, structures made of pluron/alginate-based hydrogels have been employed for direct bioprinting. It was observed that C2C12 murine myoblasts elongated after culture with 85% viability at 7 days, but a limited number of cells experienced myogenesis after 21 days (Mozetic et al., 2017).

Another field of application of 3D bioprinting is cardiac tissue engineering, which aims to create three-dimensional structures that can replicate cardiac tissue's structure and function. Cardiac muscle tissue comprises branched fibers known as cardiomyocytes, which have a single nucleus and are connected by intercalated discs. The challenge is to develop a three-dimensional construction of functional cardiac tissue with a matrix similar to the native one, a high population of cells, and adequate vascularization to guarantee the stability and contractile functionality of the tissue (Vunjak-Novakovic et al., 2010). 3D printing biomaterials can obtain cardiac tissue constructs with suitable cells; for example, Gaetani et al. (2012) used bio-printed cardiomyocyte progenitor cells with alginate constructs, and more excellent stability in alginate concentrations of 7.5% and 10% was observed, due to the higher viscosity of the bio-inks, with cell viability of 92% and 89% after 1 and 7 days of culture, respectively.

In another study by Zhu et al. (2017), a bioink of gelatin methacrylate-coated ortho nanorods combined with alginate hydrogels, cardiac fibroblasts, and cardiomyocytes was developed for 3D bioprinting of cardiac tissue. Another approach Zhang et al. (2013) used was to create constructions with perfusive hollow alginate fibers embedded in alginate gel. Jia et al. (2016) provided a 3D bioprinting vascular network using an extrusion system to print a highly organized perusable structure. For this, a cell-sensitive biolink was developed using a combination of methacrylate gelatin (GelMA), sodium alginate, and 4-arm-poly(ethylene glycol)-tetra-acrylate (PEGTA). The optimization process revealed that a GelMA concentration of 7% (w/v) showed the best cellular responses. In comparison, alginate at 3% (w/v) improved the printability of the bioink, and PEGTA at most 2% (w/v) improved the mechanical strength of the structure.

The challenge in 3D printing for bone tissue engineering is to find a material with suitable rheological properties, biocompatibility, and uniform cell incorporation capacity. In a study conducted by Bendtsen et al. (2017), a new hydrogel formulation was developed using alginate-polyvinyl alcohol (PVA) and hydroxyapatite (HA) with optimal rheological properties for 3D bioprinting of mouse calvaria MC3T3 (3T3-E1) cells on well-defined scaffolds. A systematic investigation was conducted to determine the effect of different alginate, phosphate, calcium, and PVA-HA suspension concentrations on the resulting viscosity and, thus, on the printability of the hydrogel. HA, the main mineral component of natural bone, was incorporated to promote bone formation due to its excellent osteoconductive. Degradation studies in α-MEM cell culture media showed that 3D-printed alginate-PVA-HA scaffolds remained intact for 14 days. The MC3T3 cells were evenly distributed and encapsulated in the optimal hydrogel formulation and maintained high viability throughout the 3D printing process. Therefore, this formulation of alginate-PVA-HA, which is osteoconductive, biodegradable, and can be bio-printed on three-dimensional scaffolds, is a promising candidate for treating bone defects in bone tissue engineering.

Another extracellular polymeric substance used in bioinks for 3D bioprinting is hyaluronic acid. These extracellular polymeric substances are ecological, renewable, and sustainable and have functional groups in their structures that allow chemical modification. Their formulation with chemically modified biopolymers and their combination of other components make these compounds viable with improved 3D printing properties (Pahlevanzadeh et al., 2020). On the other hand, hyaluronic acid, present in the cartilage extracellular matrix, is used to promote its formation. Bioinks based on hyaluronic acid allow stem cells to maintain their ability to differentiate and target the chondrogenic phenotype (Sakai et al., 2018). Gellan gum, a microbial polysaccharide, has also been proposed for use in cartilage tissue engineering, although the gelling temperature of gellan may affect cell viability and mechanical properties (Osmałek et al., 2014). As for blood vessel engineering, the goal is to develop structures similar to natural blood vessels. Skardal et al. (2010) used tetra polyethylene glycol and thiolated hyaluronic acid for bioprinting fibroblast-laden filaments and engineered blood vessel-like structures. After four weeks of culture, a few cases of dead cells were observed, suggesting that this 3D bioprinting method could allow the development of functional blood vessels. Ngo et al. (2020) optimized hyaluronic acid methacrylate (MeHA) for three-dimensional bioprinting to create *in vitro* test beds containing growth factors that promote regeneration in neural repair processes. Rheological studies were performed with PC-12 cells to evaluate MeHA as a bi-oink, demonstrating that it retains its shear-thinning properties during printing with and without cells. An extrusion-based 3D printer was used to bio-print various MeHA solutions combined with collagen-I, and the most optimal formulation was MeHA (10 mg/mL) with collagen-I (3 mg/mL). Schwann cell adhesion (SC) was evaluated by immunostaining β1 integrin, and it was observed that the candidate bioink favored the adhesion and migration of SCs compared to collagen-I. The MeHA/collagen-I bioink was adapted for specific neural applications by printing with neural growth factor and glial cell line-derived neurotrophic factor. The selected MeHA/collagen-I bioink was generally bioprintable, improved cellular, and favored cell adhesion, growth factor sequestration, and neural cell infiltration.

Bioprinting has also shown promise in the manufacture of structured cartilage. Mouser et al. (2020) investigated the ability to print gelatin-methacryloyl/gellan gum hydrogels (gelMA/gellan) with and without methacrylate hyaluronic acid (HAMA). These researchers demonstrated that adding HAMA improved the mechanical properties of gelMA/gellan and promoted the formation of more structured cartilage. The three-dimensional bioprinting of these hydrogels allowed the creation of cartilage constructions with high precision and reproducibility. In a recent study by Lee et al. (2021), the properties of bioink used in three-dimensional bioprinting for tissue engineering were investigated. A bioink combining hyaluronic acid and sodium alginate without chemical crosslinking agents was developed for 30% hyaluronic acid (w/v). The results showed that the combined hydrogel was successfully printed in different groups with varied viscosities at a shear rate of 0.1. The printed structures maintained their integrity after crosslinking with $CaCl_2$ and during culture periods. In addition, the proliferation rate of cells encapsulated in hyaluronic acid/sodium alginate bioink was 70% higher than in bioink containing only sodium alginate after the fourth day. These findings suggest that the 3D printable hyaluronic acid/sodium alginate hydrogel could be used as a bioink in tissue engineering applications. Three-dimensional bioprinting offers numerous possibilities in tissue engineering and regenerative medicine. Although challenges remain, such as selecting suitable biomaterials and integrating with native tissues, advances in this technology can revolutionize the field of medicine and improve people's quality of life by generating functional tissues and organs.

Conclusion

Combining extracellular polymers with bioprinting techniques has excellent potential for the manufacture of three-dimensional tissues in the future. Bioprinting is an emerging technology that uses living cells and biomaterials to create complex tissue structures. Promising results can be obtained by combining this technique with extracellular polymers, critical components of the natural extracellular matrix in tissues. One of the main advantages of combining extracellular polymers with bioprinting is that the architecture and composition of the extracellular matrix can be recreated in the printed tissue. The extracellular matrix is critical to the function and structure of tissues in the human body, providing structural support, cell signaling, and regulation of cell growth and differentiation. Incorporating extracellular polymers into bio-printed tissues can promote proper cellular interaction, and an environment conducive to tissue development and function can be created. Combining extracellular polymers with bioprinting techniques for three-dimensional tissue fabrication presents challenges and promising research areas. The selection of suitable biomaterials is critical, as careful evaluation of their properties is required to ensure compatibility with cells and the ability to promote tissue formation. In addition, achieving accurate mimicry of the natural extracellular matrix in bio-printed tissues is a significant challenge that requires a deeper understanding of its structure and function.

The integration and vascularization of printed tissues are also challenges to overcome. Integrating with surrounding tissues and forming functional blood vessels in printed tissues requires further research. In addition, the bioprinting of three-dimensional tissues at the production scale and the standardization of processes are crucial to ensuring clinical viability. A rigorous evaluation of the functionality and safety of bio-printed tissues is essential. More research is needed to develop robust assessment methods for cell viability, tissue functionality, immune response, and integration with surrounding tissues. It is also important to conduct long-term studies to assess bio-printed tissues' long-term stability and response. As for new research, combinations of extracellular polymers with other biomaterials, such as nanoparticles or bioactive peptides, can be explored to improve the properties and functionality of bio-printed tissues. Investigating printing parameters such as deposition rate and cell density and improving spatial resolution and print speed is essential to achieve more efficient and accurate bioprinting. Combining extracellular polymers with bioprinting techniques to fabricate three-dimensional tissues presents challenges that require further investigation. Selection of suitable biomaterials, extracellular matrix mimicry, integration and vascularization of printed tissues, production scale and standardization, and evaluation of functionality and safety are key focus areas. At the same time, further research can be conducted to explore combinations of biomaterials and improve printing parameters to advance the clinical application of this promising technology.

Disclaimer

None.

References

Abbasnezhad S, Biazar E, Aavani F, Kamalvand M, Heidari Keshel S, Pourjabbar B. Chemical modification of acellular fish skin as a promising biological scaffold by carbodiimide cross-linker for wound healing. *Int Wound J* (2023) 20:1566–77. doi:10.1111/iwj.14012.

Acosta S, Quintanilla-Sierra L, Mbundi L, Reboto V, Rodríguez-Cabello JC. Elastin-like recombinamers: deconstructing and recapitulating the functionality of extracellular matrix proteins using recombinant protein polymers. *Adv Funct Mater* (2020) 30:1909050. doi:10.1002/adfm.201909050.

Adepu S, Ramakrishna S. Controlled Drug Delivery Systems: Current status and future directions. *Molecules* (2021) 26:5905. doi:10.3390/molecules26195905.

Ahn G, Park JH, Kang T, Lee JW, Kang H-W, Cho D-W. Effect of pore architecture on oxygen diffusion in 3d scaffolds for tissue engineering. *J Biomech Eng* (2010) 132. doi:10.1115/1.4002429.

Akbardoost J, Amirafshari R, Mohsenzade O, Berto F. Scaling effect on the fracture toughness of bone materials using MMTS criterion. *J Mech Behav Biomed Mater* (2018) 85:72–9. doi:10.1016/j.jmbbm.2018.05.028.

Aldemir Dikici B, Reilly GC, Claeyssens F. Boosting the osteogenic and angiogenic performance of multiscale porous polycaprolactone scaffolds by *in vitro* generated extracellular matrix decoration. *ACS Appl Mater Interfaces* (2020) 12:12510–24. doi:10.1021/acsami.9b23100.

Anthon SG, Valente KP. Vascularization strategies in 3D cell culture models: from scaffold-free models to 3D bioprinting. *Int J Mol Sci* (2022) 23:14582. doi:10.3390/ijms232314582.

Avila Rodríguez M, Rodríguez Barroso L, Sánchez M. Collagen: A review on its sources and potential cosmetic applications. *J Cosmet Dermatol* (2018) 17:20–6.

Babitha S, Rachita L, Karthikeyan K, Shoba E, Janani I, Poornima B, Purna K. Electrospun protein nanofibers in healthcare: A review. *Int J Pharm* (2017) 523:52–90. doi:10.1016/j.ijpharm.2017.03.013.

Bandzerewicz A, Gadomska-Gajadhur A. Into the tissues: extracellular matrix and its artificial substitutes: cell signalling mechanisms. *Cells* (2022) 11:914. doi:10.3390/cells11050914.

Bashiri S, Ghanbarzadeh B, Ayaseh A, Dehghannya J, Ehsani A. Preparation and characterization of chitosan-coated nanostructured lipid carriers (CH-NLC) containing cinnamon essential oil for enriching milk and anti-oxidant activity. *LWT* (2020) 119:108836. doi:10.1016/j.lwt.2019.108836.

Bashor CJ, Hilton IB, Bandukwala H, Smith DM, Veiseh O. Engineering the next generation of cell-based therapeutics. *Nat Rev Drug Discov* (2022) 21:655–75. doi:10.1038/s41573-022-00476-6.

Bendtsen ST, Quinnell SP, Wei M. Development of a novel alginate-polyvinyl alcohol-hydroxyapatite hydrogel for 3D bioprinting bone tissue engineered scaffolds. *J Biomed Mater Res Part A* (2017) 105:1457–68. doi:10.1002/jbm.a.36036.

Bettinger CJ, Langer R, Borenstein JT. Engineering substrate topography at the micro- and nanoscale to control cell function. *Angew Chemie Int Ed* (2009) 48:5406–15. doi:10.1002/anie.200805179.

Bishop ES, Mostafa S, Pakvasa M, Luu HH, Lee MJ, Wolf JM, Ameer GA, He TC, Reid RR. 3-D bioprinting technologies in tissue engineering and regenerative medicine: Current and future trends. *Genes Dis* (2017) 4:185–95. doi:10.1016/j.gendis.2017.10.002.

Boni R, Ali A, Shavandi A, Clarkson AN. Current and novel polymeric biomaterials for neural tissue engineering. *J Biomed Sci* (2018) 25:90. doi:10.1186/s12929-018-0491-8.

Brovold M, Almeida JI, Pla-Palacín I, Sainz-Arnal P, Sánchez-Romero N, Rivas JJ, Almeida H, Royo-Dachary P, Serrano-Aulló T, Soker S, Baptista PM. *Naturally-derived biomaterials for tissue engineering applications,* 2018, p. 421–49. doi:10.1007/978-981-13-0947-2_23.

Bujoli B, Scimeca J-C, Verron E. Fibrin as a multipurpose physiological platform for bone tissue engineering and targeted delivery of bioactive compounds. *Pharmaceutics* (2019) 11:556. doi:10.3390/pharmaceutics11110556.

Calori IR, Braga G, de Jesus P da CC, Bi H, Tedesco AC. Polymer scaffolds as drug delivery systems. *Eur Polym J* (2020) 129:109621. doi:10.1016/j.eurpolymj.2020.109621.

Cao H, Chen M-M, Liu Y, Liu Y-Y, Huang Y-Q, Wang J-H, Chen J-D, Zhang Q-Q. Fish collagen-based scaffold containing PLGA microspheres for controlled growth factor delivery in skin tissue engineering. *Colloids Surfaces B Biointerfaces* (2015) 136:1098–106. doi:10.1016/j.colsurfb.2015.10.022.

Catoira MC, Fusaro L, Di Francesco D, Ramella M, Boccafoschi F. Overview of natural hydrogels for regenerative medicine applications. *J Mater Sci Mater Med* (2019) 30:115. doi:10.1007/s10856-019-6318-7.

Charbe NB, Tambuwala M, Palakurthi SS, Warokar A, Hromić-Jahjefendić A, Bakshi H, Zacconi F, Mishra V, Khadse S, Aljabali AA, El-Tanani M, Serrano-Aroca A, Palakurthi S. Biomedical applications of three-dimensional bioprinted craniofacial tissue engineering. *Bioeng Transl Med* (2023) 8. doi:10.1002/btm2.10333.

Chen Z-H, Ren X-L, Zhou H-H, Li X-D. The role of hyaluronic acid in biomineralization. *Front Mater Sci* (2012) 6:283–96. doi:10.1007/s11706-012-0182-4.

Cheng R, Liu L, Xiang Y, Lu Y, Deng L, Zhang H, Santos HA, Cui W. Advanced liposome-loaded scaffolds for therapeutic and tissue engineering applications. *Biomaterials* (2020) 232:119706. doi:10.1016/j.biomaterials.2019.119706.

Chernozem RV, Guselnikova O, Surmeneva MA, Postnikov PS, Abalymov AA, Parakhonskiy BV, De Roo N, Depla D, Skirtach AG, Surmenev RA. Diazonium chemistry surface treatment of piezoelectric polyhydroxybutyrate scaffolds for enhanced osteoblastic cell growth. *Appl Mater Today* (2020) 20:100758. doi:10.1016/j.apmt.2020.100758.

Chew CH, Sheu B-L, Chen A, Huang W-T, Cheng T-M, Shih C-M, Chang A, Chen C-C. Tissue-engineered vascular graft with co-culture of smooth muscle cells and human endothelial vein cells on an electrospun poly(lactic-co-glycolic acid) microtube array membrane. *Membranes (Basel)* (2021) 11:732. doi:10.3390/membranes11100732.

Collins MN, Ren G, Young K, Pina S, Reis RL, Oliveira JM. Scaffold fabrication technologies and structure/function properties in bone tissue engineering. *Adv Funct Mater* (2021) 31:2010609. doi:10.1002/adfm.202010609.

Copes F, Pien N, Van Vlierberghe S, Boccafoschi F, Mantovani D. Collagen-based tissue engineering strategies for vascular medicine. *Front Bioeng Biotechnol* (2019) 7. doi:10.3389/fbioe.2019.00166.

Cortina MS, Azar DT. Corneal angiogenesis. Encycl. Eye, Elsevier; 2010, p. 408–15. doi:10.1016/B978-0-12-374203-2.00145-7.
Dang Y, Guan J. Nanoparticle-based drug delivery systems for cancer therapy. *Smart Mater Med* (2020) 1:10–9. doi:10.1016/j.smaim.2020.04.001.
Davidson M, Ware B, Khetani S. Stem cell-derived liver cells for drug testing and disease modeling. *Discov Med* (2015) 19:349.
Davoodi E, Sarikhani E, Montazerian H, Ahadian S, Costantini M, Swieszkowski W, Willerth SM, Walus K, Mofidfar M, Toyserkani E, Khademhosseini A, Ashammakhi N. Extrusion and microfluidic-based bioprinting to fabricate biomimetic tissues and organs. *Adv Mater Technol* (2020) 5:1901044. doi:10.1002/admt.201901044.
Discher DE, Mooney DJ, Zandstra PW. Growth factors, matrices, and forces combine and control stem cells. *Science (80-)* (2009) 324:1673–7. doi:10.1126/science.1171643.
Doyle L, Wang M. Overview of Extracellular Vesicles, Their origin, composition, purpose, and methods for exosome isolation and analysis. *Cells* (2019) 8:727. doi:10.3390/cells8070727.
Dzobo K, Thomford NE, Senthebane DA, Shipanga H, Rowe A, Dandara C, Pillay M, Motaung KSC. Advances in regenerative medicine and tissue engineering: innovation and transformation of medicine. *Stem Cells Int* (2018) 2018:1–24. doi:10.1155/2018/2495848.
Ebrahimkhani MR, Young CL, Lauffenburger DA, Griffith LG, Borenstein JT. Approaches to in vitro tissue regeneration with application for human disease modeling and drug development. *Drug Discov Today* (2014) 19:754–62. doi:10.1016/j.drudis.2014.04.017.
Echave MC, Burgo LS, Pedraz JL, Orive G. Gelatin as biomaterial for tissue engineering. *Curr Pharm Des* (2017) 23. doi:10.2174/0929867324666170511123101.
Elhassan E, Devnarain N, Mohammed M, Govender T, Omolo CA. Engineering hybrid nanosystems for efficient and targeted delivery against bacterial infections. *J Control Release* (2022) 351:598–622. doi:10.1016/j.jconrel.2022.09.052.
Elkomy MH. Changing the Drug Delivery System: Does it add to non-compliance ramifications control? a simulation study on the pharmacokinetics and pharmacodynamics of atypical antipsychotic drug. *Pharmaceutics* (2020) 12:297. doi:10.3390/pharmaceutics12040297.
Eming SA, Krieg T, Davidson JM. Inflammation in wound repair: molecular and cellular mechanisms. *J Invest Dermatol* (2007) 127:514–25. doi:10.1038/sj.jid.5700701.
Fedorovich NE, Schuurman W, Wijnberg HM, Prins H-J, van Weeren PR, Malda J, Alblas J, Dhert WJA. Biofabrication of osteochondral tissue equivalents by printing topologically defined, cell-laden hydrogel scaffolds. *Tissue Eng Part C Methods* (2012) 18:33–44. doi:10.1089/ten.tec.2011.0060.
Feng, Z, Fan H, Cheng L, Zhang H, Fan H, Liu J. Advanced biomimetic nanomaterials for non-invasive disease diagnosis. *Front Mater* (2021) 8:664795.
Fereshteh Z. Freeze-drying technologies for 3D scaffold engineering. Funct. 3D Tissue Eng. Scaffolds, Elsevier; 2018, p. 151–74. doi:10.1016/B978-0-08-100979-6.00007-0.
Frantz C, Stewart KM, Weaver VM. The extracellular matrix at a glance. *J Cell Sci* (2010) 123:4195–200. doi:10.1242/jcs.023820.

Fujimaki H, Matsumine H, Osaki H, Ueta Y, Kamei W, Shimizu M, Hashimoto K, Fujii K, Kazama T, Matsumoto T, Niimi Y, Miyata M, Sakurai H. Dedifferentiated fat cells in polyglycolic acid-collagen nerve conduits promote rat facial nerve regeneration. *Regen Ther* (2019) 11:240–8. doi:10.1016/j.reth.2019.08.004.

Furtado M, Chen L, Chen Z, Chen A, Cui W. Development of fish collagen in tissue regeneration and drug delivery. *Eng Regen* (2022) 3:217–31. doi:10.1016/j.engreg. 2022.05.002.

Gaetani R, Doevendans PA, Metz CHG, Alblas J, Messina E, Giacomello A, Sluijter JPG. Cardiac tissue engineering using tissue printing technology and human cardiac progenitor cells. *Biomaterials* (2012) 33:1782–90. doi:10.1016/j.biomaterials. 2011.11.003.

Gaihre B, Potes MA, Serdiuk V, Tilton M, Liu X, Lu L. Two-dimensional nanomaterials-added dynamism in 3D printing and bioprinting of biomedical platforms: Unique opportunities and challenges. *Biomaterials* (2022) 284:121507. doi:10.1016/ j.biomaterials.2022.121507.

Garreta E, Oria R, Tarantino C, Pla-Roca M, Prado P, Fernández-Avilés F, Campistol JM, Samitier J, Montserrat N. Tissue engineering by decellularization and 3D bioprinting. *Mater Today* (2017) 20:166–78. doi:10.1016/j.mattod.2016.12.005.

Gresham RCH, Bahney CS, Leach JK. Growth factor delivery using extracellular matrix-mimicking substrates for musculoskeletal tissue engineering and repair. *Bioact Mater* (2021) 6:1945–56. doi:10.1016/j.bioactmat.2020.12.012.

Grimaudo MA, Krishnakumar GS, Giusto E, Furlani F, Bassi G, Rossi A, Molinari F, Lista F, Montesi M, Panseri S. Bioactive injectable hydrogels for on demand molecule/cell delivery and for tissue regeneration in the central nervous system. *Acta Biomater* (2022) 140:88–101. doi:10.1016/j.actbio.2021.11.038.

Guilak F, Cohen DM, Estes BT, Gimble JM, Liedtke W, Chen CS. Control of stem cell fate by physical interactions with the extracellular matrix. *Cell Stem Cell* (2009) 5:17–26. doi:10.1016/j.stem.2009.06.016.

Guimarães CF, Gasperini L, Marques AP, Reis RL. The stiffness of living tissues and its implications for tissue engineering. *Nat Rev Mater* (2020) 5:351–70. doi:10.1038/s41578-019-0169-1.

Guo Z, Poot AA, Grijpma DW. Advanced polymer-based composites and structures for biomedical applications. *Eur Polym J* (2021) 149:110388. doi:10.1016/j.eurpolymj. 2021.110388.

Han F, Wang J, Ding L, Hu Y, Li W, Yuan Z, Guo Q, Zhu C, Yu L, Wang H, Zhao Z, Jia L, Li J, Yu Y, Zhang W, Chu G, Chen S, Li B. Tissue engineering and regenerative medicine: achievements, future, and sustainability in Asia. *Front Bioeng Biotechnol* (2020) 8. doi:10.3389/fbioe.2020.00083.

Han L, Xu J, Lu X, Gan D, Wang Z, Wang K, Zhang H, Yuan H, Weng J. Biohybrid methacrylated gelatin/polyacrylamide hydrogels for cartilage repair. *J Mater Chem B* (2017) 5:731–41. doi:10.1039/C6TB02348G.

Hao D, Swindell HS, Ramasubramanian L, Liu R, Lam KS, Farmer DL, Wang A. Extracellular matrix mimicking nanofibrous scaffolds modified with mesenchymal stem cell-derived extracellular vesicles for improved vascularization. *Front Bioeng Biotechnol* (2020) 8. doi:10.3389/fbioe.2020.00633.

He W, Reaume M, Hennenfent M, Lee BP, Rajachar R. Biomimetic hydrogels with spatial- and temporal-controlled chemical cues for tissue engineering. *Biomater Sci* (2020) 8:3248–69. doi:10.1039/D0BM00263A.

Hoshiba T, Lu H, Kawazoe N, Chen G. Decellularized matrices for tissue engineering. *Expert Opin Biol Ther* (2010) 10:1717–28. doi:10.1517/14712598.2010.534079.

Huang C, Dai J, Zhang XA. Environmental physical cues determine the lineage specification of mesenchymal stem cells. *Biochim Biophys Acta - Gen Subj* (2015) 1850:1261–6. doi:10.1016/j.bbagen.2015.02.011.

Humaira, Raza Bukhari SA, Shakir HA, Khan M, Saeed S, Ahmad I, Muzammil K, Franco M, Irfan M, Li K. Hyaluronic acid-based nanofibers: Electrospun synthesis and their medical applications; recent developments and future perspective. *Front Chem* (2022) 10. doi:10.3389/fchem.2022.1092123.

Hunt NC, Hallam D, Karimi A, Mellough CB, Chen J, Steel DHW, Lako M. 3D culture of human pluripotent stem cells in RGD-alginate hydrogel improves retinal tissue development. *Acta Biomater* (2017) 49:329–43. doi:10.1016/j.actbio.2016.11.016.

Hutmacher DW. Scaffolds in tissue engineering bone and cartilage. *Biomaterials* (2000) 21:2529–43. doi:10.1016/S0142-9612(00)00121-6.

Hynes RO. The evolution of metazoan extracellular matrix. *J Cell Biol* (2012) 196:671–9. doi:10.1083/jcb.201109041.

Hynes RO. The Extracellular matrix: not just pretty fibrils. *Science (80-)* (2009) 326:1216–9. doi:10.1126/science.1176009.

Iacob A-T, Drăgan M, Ionescu O-M, Profire L, Ficai A, Andronescu E, Confederat LG, Lupascu D. An Overview of Biopolymeric Electrospun Nanofibers Based on Polysaccharides for Wound Healing Management. *Pharmaceutics* (2020) 12:983. doi:10.3390/
pharmaceutics12100983.

Idini M, Wieringa P, Rocchiccioli S, Nieddu G, Ucciferri N, Formato M, Lepedda A, Moroni L. Glycosaminoglycan functionalization of electrospun scaffolds enhances Schwann cell activity. *Acta Biomater* (2019) 96:188–202. doi:10.1016/j.actbio. 2019.06.054.

Jain KK. An Overview of drug delivery systems, 2020, p. 1–54. doi:10.1007/978-1-4939-9798-5_1.

Janarthanan G, Noh I. Recent trends in metal ion based hydrogel biomaterials for tissue engineering and other biomedical applications. *J Mater Sci Technol* (2021) 63:35–53. doi:10.1016/j.jmst.2020.02.052.

Jartarkar SR, Patil A, Wollina U, Gold MH, Stege H, Grabbe S, Goldust M. New diagnostic and imaging technologies in dermatology. *J Cosmet Dermatol* (2021) 20:3782–7. doi:10.1111/jocd.14499.

Jeon O, Lee YB, Hinton TJ, Feinberg AW, Alsberg E. Cryopreserved cell-laden alginate microgel bioink for 3D bioprinting of living tissues. *Mater Today Chem* (2019) 12:61–70. doi:10.1016/j.mtchem.2018.11.009.

Jia W, Gungor-Ozkerim PS, Zhang YS, Yue K, Zhu K, Liu W, Pi Q, Byambaa B, Dokmeci MR, Shin SR, Khademhosseini, A. Direct 3D bioprinting of perfusable vascular constructs using a blend bioink. *Biomaterials* (2016) 106:58–68. doi:10.1016/j.biomaterials.2016.07.038.

Jing X, Hu X, Feng P, Liu Y, Yang J. Modification of nanofibrous scaffolds to mimic extracellular matrix in physical and chemical structuring. *Polym Eng Sci* (2023) 63:467–77. doi:10.1002/pen.26221.

Johnson PJ, Tatara A, Shiu A, Sakiyama-Elbert SE. Controlled release of neurotrophin-3 and platelet-derived growth factor from fibrin scaffolds containing neural progenitor cells enhances survival and differentiation into neurons in a subacute model of SCI. *Cell Transplant* (2010) 19:89–101. doi:10.3727/096368909X477273.

Jorfi M, Foster EJ. Recent advances in nanocellulose for biomedical applications. *J Appl Polym Sci* (2015) 132:n/a-n/a. doi:10.1002/app.41719.

Kabu S, Gao Y, Kwon BK, Labhasetwar V. Drug delivery, cell-based therapies, and tissue engineering approaches for spinal cord injury. *J Control Release* (2015) 219:141–54. doi:10.1016/j.jconrel.2015.08.060.

Kakinoki S, Nishioka S, Arichi Y, Yamaoka T. Stable and direct coating of fibronectin-derived Leu-Asp-Val peptide on ePTFE using one-pot tyrosine oxidation for endothelial cell adhesion. *Colloids Surfaces B Biointerfaces* (2022) 216:112576. doi:10.1016/j.colsurfb.2022.112576.

Kang H-W, Lee SJ, Ko IK, Kengla C, Yoo JJ, Atala A. A 3D bioprinting system to produce human-scale tissue constructs with structural integrity. *Nat Biotechnol* (2016) 34:312–9. doi:10.1038/nbt.3413.

Karamanos NK, Theocharis AD, Neill T, Iozzo R V. Matrix modeling and remodeling: A biological interplay regulating tissue homeostasis and diseases. *Matrix Biol* (2019) 75–76:1–11. doi:10.1016/j.matbio.2018.08.007.

Karsdal MA, Nielsen MJ, Sand JM, Henriksen K, Genovese F, Bay-Jensen A-C, Smith V, Adamkewicz JI, Christiansen C, Leeming DJ. Extracellular matrix remodeling: the common denominator in connective tissue diseases possibilities for evaluation and current understanding of the matrix as more than a passive architecture, but a key player in tissue failure. *Assay Drug Dev Technol* (2013) 11:70–92. doi:10.1089/adt.2012.474.

Khanna A, Zamani M, Huang NF. Extracellular matrix-based biomaterials for cardiovascular tissue engineering. *J Cardiovasc Dev Dis* (2021) 8:137. doi:10.3390/jcdd8110137.

Koivisto JT, Gering C, Karvinen J, Maria Cherian R, Belay B, Hyttinen J, Aalto-Setala K, Kellomaki M, Parraga J. Mechanically biomimetic gelatin–gellan gum hydrogels for 3D culture of beating human cardiomyocytes. *ACS Appl Mater Interfaces* (2019) 11:20589–602. doi:10.1021/acsami.8b22343.

Krishnan V, Peng K, Sarode A, Prakash S, Zhao Z, Filippov SK, Todorova K, Sell BR, Lujano O, Bakre S, Pusuluri A, Vogus D, Tsai KY, Mandinova A, Mitragotri S. Hyaluronic acid conjugates for topical treatment of skin cancer lesions. *Sci Adv* (2021) 7. doi:10.1126/sciadv.abe6627.

Kular JK, Basu S, Sharma RI. The extracellular matrix: Structure, composition, age-related differences, tools for analysis and applications for tissue engineering. *J Tissue Eng* (2014) 5:204173141455711. doi:10.1177/2041731414557112.

Kumar V, Taylor N, Jalan A, Hwang L, Wang B, Hartgerink J. A nanostructured synthetic collagen mimic for hemostasis. *Biomacromolecules* (2014) 15:1584–90.

Lancaster MA, Huch M. Disease modelling in human organoids. *Dis Model Mech* (2019) 12. doi:10.1242/dmm.039347.

Lavdas AA, Matsas R. Schwann Cell Morphology☆. Ref. Modul. Biomed. Sci., Elsevier; 2014. doi:10.1016/B978-0-12-801238-3.04758-9.

Ledford H. Translational research: 4 ways to fix the clinical trial. *Nature* (2011) 477:526–8. doi:10.1038/477526a.

Lee J, Manoharan V, Cheung L, Lee S, Cha B-H, Newman P, Farzad R, Mehrotra S, Zhang K, Khan F, Ghaderi M, Lin Y-D, Li Y-D, Aftab S, Mostafalu P, Miscuglio M, Li J, Mandal BB, Hussain MA, Wan K-T, Tang X, Khademshosseini A, Shin SR. Nanoparticle-based hybrid scaffolds for deciphering the role of multimodal cues in cardiac tissue engineering. *ACS Nano* (2019) 13:12525–39. doi:10.1021/acsnano.9b03050.

Lee SJ, Seok JM, Lee JH, Lee J, Kim WD, Park SA. Three-dimensional printable hydrogel using a hyaluronic acid/sodium alginate bio-ink. *Polymers (Basel)* (2021) 13:794. doi:10.3390/polym13050794.

Li G, Wang L, Pan W, Yang F, Jiang W, Wu X, Kong X, Dai K, Hao Y. *In vitro* and *in vivo* study of additive manufactured porous Ti6Al4V scaffolds for repairing bone defects. *Sci Rep* (2016) 6:34072. doi:10.1038/srep34072.

Li R, Liu K, Huang X, Li D, Ding J, Liu B, Chen X. Bioactive materials promote wound healing through modulation of cell behaviors. *Adv Sci* (2022) 9:2105152. doi:10.1002/advs.202105152.

Li S, Niu D, Shi T, Yun W, Yan S, Xu G, Yin J. Injectable, *in situ* self-cross-linking, self-healing poly(<scp>l</scp>-glutamic acid)/polyethylene glycol hydrogels for cartilage tissue engineering. *ACS Biomater Sci Eng* (2023) 9:2625–35. doi:10.1021/acsbiomaterials.3c00041.

Li Y, Rodrigues J, Tomás H. Injectable and biodegradable hydrogels: gelation, biodegradation and biomedical applications. *Chem Soc Rev* (2012) 41:2193–221. doi:10.1039/C1CS15203C.

Lim Y-S, Ok Y-J, Hwang S-Y, Kwak J-Y, Yoon S. Marine Collagen as a promising biomaterial for biomedical applications. *Mar Drugs* (2019) 17:467. doi:10.3390/md17080467.

Liu S, Cao S, Guo J, Luo L, Zhou Y, Lin C, Shi J, Fan C, Lv M, Wang L. Graphene oxide–silver nanocomposites modulate biofilm formation and extracellular polymeric substance (EPS) production. *Nanoscale* (2018) 10:19603–11. doi:10.1039/C8NR04064H.

Liu S, Wu G, Chen X, Zhang X, Yu J, Liu M, Zhang Y, Wang P. Degradation behavior *in vitro* of carbon nanotubes (CNTs)/Poly(lactic acid) (PLA) Composite Suture. *Polymers (Basel)* (2019) 11:1015. doi:10.3390/polym11061015.

Long K, Liu Y, Li W, Wang L, Liu S, Wang Y, Wang Z, Ren L. Improving the mechanical properties of collagen-based membranes using silk fibroin for corneal tissue engineering. *J Biomed Mater Res Part A* (2015) 103:1159–68. doi:10.1002/jbm.a.35268.

Lou T, Yan X, Wang X. Chitosan coated polyacrylonitrile nanofibrous mat for dye adsorption. *Int J Biol Macromol* (2019) 135:919–25. doi:10.1016/j.ijbiomac.2019.06.008.

Lu P, Takai K, Weaver VM, Werb Z. Extracellular matrix degradation and remodeling in development and disease. *Cold Spring Harb Perspect Biol* (2011) 3:a005058–a005058. doi:10.1101/cshperspect.a005058.

Lukin I, Erezuma I, Maeso L, Zarate J, Desimone MF, Al-Tel TH, Dolastshahi-Pirouz A, Orive G. Progress in gelatin as biomaterial for tissue engineering. *Pharmaceutics* (2022) 14:1177. doi:10.3390/pharmaceutics14061177.

Luo Y, Lin X, Huang P. 3D Bioprinting of artificial tissues: construction of biomimetic microstructures. *Macromol Biosci* (2018) 18:1800034. doi:10.1002/mabi.201800034.

Lyu Y, Xie J, Liu Y, Xiao M, Li Y, Yang J, Yang J, Liu W. Injectable hyaluronic acid hydrogel loaded with functionalized human mesenchymal stem cell aggregates for repairing infarcted myocardium. *ACS Biomater Sci Eng* (2020) 6:6926–37. doi:10.1021/acsbiomaterials.0c01344.

Madni A, Kousar R, Naeem N, Wahid F. Recent advancements in applications of chitosan-based biomaterials for skin tissue engineering. *J Bioresour Bioprod* (2021) 6:11–25. doi:10.1016/j.jobab.2021.01.002.

Malda J, Visser J, Melchels FP, Jüngst T, Hennink WE, Dhert WJA, Hutmacher DW. 25th Anniversary article: engineering hydrogels for biofabrication. *Adv Mater* (2013) 25:5011–28. doi:10.1002/adma.201302042.

Malone ER, Oliva M, Sabatini PJB, Stockley TL, Siu LL. Molecular profiling for precision cancer therapies. *Genome Med* (2020) 12:8. doi:10.1186/s13073-019-0703-1.

Man W, Yang S, Cao Z, Lu J, Kong X, Sun X, Zhao L, Guo Y, Yao S, Wang G, Wang X. A multi-modal delivery strategy for spinal cord regeneration using a composite hydrogel presenting biophysical and biochemical cues synergistically. *Biomaterials* (2021) 276:120971. doi:10.1016/j.biomaterials.2021.120971.

Mao Y, Schwarzbauer JE. Fibronectin fibrillogenesis, a cell-mediated matrix assembly process. *Matrix Biol* (2005) 24:389–99. doi:10.1016/j.matbio.2005.06.008.

Martin JD, Cabral H, Stylianopoulos T, Jain RK. Improving cancer immunotherapy using nanomedicines: progress, opportunities and challenges. *Nat Rev Clin Oncol* (2020) 17:251–66. doi:10.1038/s41571-019-0308-z.

Mecham RP. Elastin in lung development and disease pathogenesis. *Matrix Biol* (2018) 73:6–20. doi:10.1016/j.matbio.2018.01.005.

Meng W, He C, Hao Y, Wang L, Li L, Zhu G. Prospects and challenges of extracellular vesicle-based drug delivery system: considering cell source. *Drug Deliv* (2020) 27:585–98. doi:10.1080/10717544.2020.1748758.

Midwood KS, Chiquet M, Tucker RP, Orend G. Tenascin-C at a glance. *J Cell Sci* (2016). doi:10.1242/jcs.190546.

Mihardja SS, Sievers RE, Lee RJ. The effect of polypyrrole on arteriogenesis in an acute rat infarct model. *Biomaterials* (2008) 29:4205–10. doi:10.1016/j.biomaterials.2008.07.021.

Mouser VHM, Levato R, Mensinga A, Dhert WJA, Gawlitta D, Malda J. Bio-ink development for three-dimensional bioprinting of hetero-cellular cartilage constructs. *Connect Tissue Res* (2020) 61:137–51. doi:10.1080/03008207.2018.1553960.

Mozetic P, Giannitelli SM, Gori M, Trombetta M, Rainer A. Engineering muscle cell alignment through 3D bioprinting. *J Biomed Mater Res Part A* (2017) 105:2582–8. doi:10.1002/jbm.a.36117.

Mushtaq F, Raza ZA, Batool SR, Zahid M, Onder OC, Rafique A, Nazeer MA. Preparation, properties, and applications of gelatin-based hydrogels (GHs) in the environmental, technological, and biomedical sectors. *Int J Biol Macromol* (2022) 218:601–33. doi:10.1016/j.ijbiomac.2022.07.168.

Muttenthaler M, King GF, Adams DJ, Alewood PF. Trends in peptide drug discovery. *Nat Rev Drug Discov* (2021) 20:309–25. doi:10.1038/s41573-020-00135-8.

Naderi H, Matin MM, Bahrami AR. Review paper: critical issues in tissue engineering: biomaterials, cell sources, angiogenesis, and drug delivery systems. *J Biomater Appl* (2011) 26:383–417. doi:10.1177/0885328211408946.

Ngo TB, Spearman BS, Hlavac N, Schmidt CE. Three-dimensional bioprinted hyaluronic acid hydrogel test beds for assessing neural cell responses to competitive growth stimuli. *ACS Biomater Sci Eng* (2020) 6:6819–30. doi:10.1021/acsbiomaterials.0c00940.

Nguyen DT, McCanless JD, Mecwan MM, Noblett AP, Haggard WO, Smith RA, Bumgardner JD. Balancing mechanical strength with bioactivity in chitosan–calcium phosphate 3D microsphere scaffolds for bone tissue engineering: air- vs. freeze-drying processes. *J Biomater Sci Polym Ed* (2013) 24:1071–83. doi:10.1080/09205063.2012.735099.

Ni P, Ye S, Xiong S, Zhong M, Shan J, Yuan T, Liang J, Fan Y, Zhang X. A chitosan-optimized polyethyleneimine/polyacrylic acid multifunctional hydrogel for reducing the risk of ulcerative arterial bleeding. *J Mater Chem B* (2023) 11:5207–22. doi:10.1039/D3TB00239J.

Niemczyk-Soczynska B, Gradys A, Sajkiewicz P. Hydrophilic surface functionalization of electrospun nanofibrous scaffolds in tissue engineering. *Polymers (Basel)* (2020) 12:2636. doi:10.3390/polym12112636.

Nosrati H, Barzegari P, Danafar H, Kheiri Manjili H. Biotin-functionalized copolymeric PEG-PCL micelles for in vivo tumour-targeted delivery of artemisinin. *Artif Cells, Nanomedicine, Biotechnol* (2019) 47:104–14. doi:10.1080/21691401.2018.1543199.

O'Brien CM, Holmes B, Faucett S, Zhang LG. Three-dimensional printing of nanomaterial scaffolds for complex tissue regeneration. *Tissue Eng Part B Rev* (2015) 21:103–14. doi:10.1089/ten.teb.2014.0168.

O'Brien FJ. Biomaterials & scaffolds for tissue engineering. *Mater Today* (2011) 14:88–95. doi:10.1016/S1369-7021(11)70058-X.

Oliva F, Marsilio E, Asparago G, Frizziero A, Berardi AC, Maffulli N. The impact of hyaluronic acid on tendon physiology and its clinical application in tendinopathies. *Cells* (2021) 10:3081. doi:10.3390/cells10113081.

Orive G, Santos E, Pedraz JL, Hernández RM. Application of cell encapsulation for controlled delivery of biological therapeutics. *Adv Drug Deliv Rev* (2014) 67–68:3–14. doi:10.1016/j.addr.2013.07.009.

Osmałek T, Froelich A, Tasarek S. Application of gellan gum in pharmacy and medicine. *Int J Pharm* (2014) 466:328–40. doi:10.1016/j.ijpharm.2014.03.038.

Osmani RAM, Singh E, Jadhav K, Jadhav S, Banerjee R. Biopolymers and biocomposites: Nature's tools for wound healing and tissue engineering. Appl. Adv. Green Mater., Elsevier; 2021, p. 573–630. doi:10.1016/B978-0-12-820484-9.00023-4.

Ottani V, Raspanti M, Ruggeri A. Collagen structure and functional implications. *Micron* (2001) 32:251–60. doi:10.1016/S0968-4328(00)00042-1.

Pahlevanzadeh F, Emadi R, Valiani A, Kharaziha M, Poursamar SA, Bakhsheshi-Rad HR, Ismail AF, RamaKrishna S, Berto F. Three-dimensional printing constructs based on the chitosan for tissue regeneration: state of the art, developing directions and prospect trends. *Materials (Basel)* (2020) 13:2663. doi:10.3390/ma13112663.

Parenteau-Bareil R, Gauvin R, Berthod F. Collagen-based biomaterials for tissue engineering applications. *Materials (Basel)* (2010) 3:1863–87. doi:10.3390/ma3031863.

Park J-R, Van Guyse JFR, Podevyn A, Bolle ECL, Bock N, Linde E, Celina M, Hoogenboom R, Dargaville TR. Influence of side-chain length on long-term release kinetics from poly(2-oxazoline)-drug conjugate networks. *Eur Polym J* (2019) 120:109217. doi:10.1016/j.eurpolymj.2019.109217.

Patel R, Santhosh M, Dash JK, Karpoormath R, Jha A, Kwak J, Patel M, Kim JH. Ile-Lys-Val-ala-Val (IKVAV) peptide for neuronal tissue engineering. *Polym Adv Technol* (2019) 30:4–12. doi:10.1002/pat.4442.

Peng X, Xia X, Xu X, Yang X, Yang B, Zhao P, Yuan W, Yan-Chiu PW, Bian L. Ultrafast self-gelling powder mediates robust wet adhesion to promote healing of gastrointestinal perforations. *Sci Adv* (2021) 7. doi:10.1126/sciadv.abe8739.

Peppas NA, Hilt JZ, Khademhosseini A, Langer R. Hydrogels in biology and medicine: from molecular principles to bionanotechnology. *Adv Mater* (2006) 18:1345–60. doi:10.1002/adma.200501612.

Pina S, Ribeiro VP, Marques CF, Maia FR, Silva TH, Reis RL, Oliveira JM. Scaffolding strategies for tissue engineering and regenerative medicine applications. *Materials (Basel)* (2019) 12:1824. doi:10.3390/ma12111824.

Platt VM, Szoka FC. Anticancer therapeutics: targeting macromolecules and nanocarriers to hyaluronan or CD44, a Hyaluronan Receptor. *Mol Pharm* (2008) 5:474–86. doi:10.1021/mp800024g.

Poole JJA, Mostaço-Guidolin LB. Optical microscopy and the extracellular matrix structure: a review. *Cells* (2021) 10:1760. doi:10.3390/cells10071760.

Presta G, Puliatti A, Bonetti L, Tolotti A, Sari D, Valcarenghi D. Effectiveness of hyaluronic acid gel (Jalosome soothing gel) for the treatment of radiodermatitis in a patient receiving head and neck radiotherapy associated with cetuximab: A case report and review. *Int Wound J* (2019) 16:1433–9. doi:10.1111/iwj.13210.

Qin X, Yu C, Wei J, Li L, Zhang C, Wu Q, Liu J, Yao SQ. Rational design of nanocarriers for intracellular protein delivery. *Adv Mater* (2019) 31:1902791. doi:10.1002/adma.201902791.

Qiu J, Li J, Wang G, Zheng L, Ren N, Liu H, Wei T, Huaidong J, Yingjun W. *In vitro* investigation on the biodegradability and biocompatibility of genipin cross-linked porcine acellular dermal matrix with intrinsic fluorescence. *ACS Appl Mater Interfaces* (2013) 5:344–50. doi:10.1021/am302272k.

Ravichandran R, Sundarrajan S, Venugopal JR, Mukherjee S, Ramakrishna S. Advances in polymeric systems for tissue engineering and biomedical applications. *Macromol Biosci* (2012) 12:286–311. doi:10.1002/mabi.201100325.

Ravindran Girija A, Palaninathan V, Strudwick X, Balasubramanian S, Dasappan Nair S, Cowin AJ. Collagen-functionalized electrospun smooth and porous polymeric scaffolds for the development of human skin-equivalent. *RSC Adv* (2020) 10:26594–603. doi:10.1039/D0RA04648E.

Raynald, Shu B, Liu X, Zhou J, Huang H, Wang J, Sun X, Qin C, An Y. Polypyrrole/polylactic acid nanofibrous scaffold cotransplanted with bone marrow stromal cells promotes the functional recovery of spinal cord injury in rats. *CNS Neurosci Ther* (2019) 25:951–64. doi:10.1111/cns.13135.

Reis R, Neves N, Mano J, Gomes M, Marques A, Azevedo H. *Natural-based polymers for biomedical applications*. Elsevier; 2008.

Rezvani Ghomi E, Nourbakhsh N, Akbari Kenari M, Zare M, Ramakrishna S. Collagen-based biomaterials for biomedical applications. *J Biomed Mater Res Part B Appl Biomater* (2021a) 109:1986–99.

Rezvani Ghomi E, Nourbakhsh N, Akbari Kenari M, Zare M, Ramakrishna S. Collagen-based biomaterials for biomedical applications. *J Biomed Mater Res Part B Appl Biomater* (2021b) 109:1986–99. doi:10.1002/jbm.b.34881.

Ricard-Blum S. The Collagen Family. *Cold Spring Harb Perspect Biol* (2011) 3:a004978–a004978. doi:10.1101/cshperspect.a004978.

Rodrigues LR, Teixeira JA, Schmitt FL, Paulsson M, Lindmark-Mänsson H. The Role of osteopontin in tumor progression and metastasis in breast cancer. *Cancer Epidemiol Biomarkers Prev* (2007) 16:1087–97. doi:10.1158/1055-9965.EPI-06-1008.

Rosellini E, Barbani N, Lazzeri L, Cascone MG. Biomimetic and bioactive small diameter tubular scaffolds for vascular tissue engineering. *Biomimetics* (2022) 7:199. doi:10.3390/biomimetics7040199.

Sachlos E, Czernuszka J. Making tissue engineering scaffolds work. review: the application of solid freeform fabrication technology to the production of tissue engineering scaffolds. *Eur Cells Mater* (2003) 5:29–40. doi:10.22203/eCM.v005a03.

Sahana TG, Rekha PD. Biopolymers: Applications in wound healing and skin tissue engineering. *Mol Biol Rep* (2018) 45:2857–67. doi:10.1007/s11033-018-4296-3.

Sahoo DR, Biswal T. Alginate and its application to tissue engineering. *SN Appl Sci* (2021) 3:30. doi:10.1007/s42452-020-04096-w.

Sakai S, Ohi H, Hotta T, Kamei H, Taya M. Differentiation potential of human adipose stem cells bioprinted with hyaluronic acid/gelatin-based bioink through microextrusion and visible light-initiated crosslinking. *Biopolymers* (2018) 109:e23080. doi:10.1002/bip.23080.

Savaskan Yilmaz S, Yildirim N, Misir M, Misirlioglu Y, Celik E. Synthesis, characterization of a new polyacrylic acid superabsorbent, some heavy metal ion sorption, the adsorption isotherms, and quantum chemical investigation. *Materials (Basel)* (2020) 13:4390. doi:10.3390/ma13194390.

Sayyar S, Murray E, Thompson BC, Chung J, Officer DL, Gambhir S, Spinks GM, Wallace GG. Processable conducting graphene/chitosan hydrogels for tissue engineering. *J Mater Chem B* (2015) 3:481–90. doi:10.1039/C4TB01636J.

Sgambato A, Cipolla L, Russo L. Bioresponsive Hydrogels: chemical strategies and perspectives in tissue engineering. *Gels* (2016) 2:28. doi:10.3390/gels2040028.

Sherman VR, Yang W, Meyers MA. The materials science of collagen. *J Mech Behav Biomed Mater* (2015) 52:22–50. doi:10.1016/j.jmbbm.2015.05.023.

Shin YJ, Shafranek RT, Tsui JH, Walcott J, Nelson A, Kim D-H. 3D bioprinting of mechanically tuned bioinks derived from cardiac decellularized extracellular matrix. *Acta Biomater* (2021) 119:75–88. doi:10.1016/j.actbio.2020.11.006.

Shoichet MS. Polymer scaffolds for biomaterials applications. *Macromolecules* (2010) 43:581–91. doi:10.1021/ma901530r.

Shoulders MD, Raines RT. Collagen structure and stability. *Annu Rev Biochem* (2009) 78:929–58. doi:10.1146/annurev.biochem.77.032207.120833.

Skardal A, Zhang J, Prestwich GD. Bioprinting vessel-like constructs using hyaluronan hydrogels crosslinked with tetrahedral polyethylene glycol tetracrylates. *Biomaterials* (2010) 31:6173–81. doi:10.1016/j.biomaterials.2010.04.045.

Song R, Murphy M, Li C, Ting K, Soo C, Zheng Z. Current development of biodegradable polymeric materials for biomedical applications. *Drug Des Devel Ther* (2018):3117–45.

Sood A, Gupta A, Agrawal G. Recent advances in polysaccharides based biomaterials for drug delivery and tissue engineering applications. *Carbohydr Polym Technol Appl* (2021) 2:100067. doi:10.1016/j.carpta.2021.100067.

Soriano-Ruiz JL, Calpena-Campmany AC, Silva-Abreu M, Halbout-Bellowa L, Bozal-de Febrer N, Rodríguez-Lagunas MJ, Clares-Naverps B. Design and evaluation of a multifunctional thermosensitive poloxamer-chitosan-hyaluronic acid gel for the treatment of skin burns. *Int J Biol Macromol* (2020) 142:412–22. doi:10.1016/j.ijbiomac.2019.09.113.

de Sousa Victor R, Marcelo da Cunha Santos A, Viana de Sousa B, de Araújo Neves, G, Navarro de Lima Santana L, Rodrigues Menezes R. A review on Chitosan's uses as biomaterial: Tissue engineering, drug delivery systems and cancer treatment. *Materials (Basel)* (2020) 13:4995.

Srivastava A, Yadav T, Sharma S, Nayak A, Akanksha Kumari A, Mishra N. Polymers in Drug Delivery. *J Biosci Med* (2016) 04:69–84. doi:10.4236/jbm.2016.41009.

Sun C-P, Lan H-R, Fang X-L, Yang X-Y, Jin K-T. Organoid models for precision cancer immunotherapy. *Front Immunol* (2022) 13. doi:10.3389/fimmu.2022.770465.

Sun W, Liu W, Wu Z, Chen H. Chemical surface modification of polymeric biomaterials for biomedical applications. *Macromol Rapid Commun* (2020) 41:1900430. doi:10.1002/marc.201900430.

Suri S, Han L-H, Zhang W, Singh A, Chen S, Schmidt CE. Solid freeform fabrication of designer scaffolds of hyaluronic acid for nerve tissue engineering. *Biomed Microdevices* (2011) 13:983–93. doi:10.1007/s10544-011-9568-9.

Tan K-X, Chamundeswari VN, Loo SCJ. Prospects of kefiran as a food-derived biopolymer for agri-food and biomedical applications. *RSC Adv* (2020) 10:25339–51. doi:10.1039/D0RA02810J.

Tan Y, Ma L, Chen X, Ran Y, Tong Q, Tang L, Li X. Injectable hyaluronic acid/hydroxyapatite composite hydrogels as cell carriers for bone repair. *Int J Biol Macromol* (2022) 216:547–57. doi:10.1016/j.ijbiomac.2022.07.009.

Taylor SJ, Rosenzweig ES, McDonald JW, Sakiyama-Elbert SE. Delivery of neurotrophin-3 from fibrin enhances neuronal fiber sprouting after spinal cord injury. *J Control Release* (2006) 113:226–35. doi:10.1016/j.jconrel.2006.05.005.

Tombling BJ, Zhang Y, Huang Y-H, Craik DJ, Wang CK. The emerging landscape of peptide-based inhibitors of PCSK9. *Atherosclerosis* (2021) 330:52–60. doi:10.1016/j.atherosclerosis.2021.06.903.

Tsur-Gang O, Ruvino E, Landa N, Holbova R, Feinberg M, Leor J, Cohen S. The effects of peptide-based modification of alginate on left ventricular remodeling and function after myocardial infarction. *Biomaterials* (2009) 30:189–95. doi:10.1016/j.biomaterials.2008.09.018.

Vacanti JP, Langer R. Tissue engineering: the design and fabrication of living replacement devices for surgical reconstruction and transplantation. *Lancet* (1999) 354:S32–4. doi:10.1016/S0140-6736(99)90247-7.

Vismara I, Papa S, Veneruso V, Mauri E, Mariani A, De Paola M, Affatato R, Rossetti A, Sponchioni M, Moscatelli D, Sacchetti A, Rossi F, Forloni G, Veglianese P. Selective modulation of a1 astrocytes by drug-loaded nano-structured gel in spinal cord injury. *ACS Nano* (2020) 14:360–71. doi:10.1021/acsnano.9b05579.

Vunjak-Novakovic G, Tandon N, Godier A, Maidhof R, Marsano A, Martens TP, Radisic M. Challenges in cardiac tissue engineering. *Tissue Eng Part B Rev* (2010) 16:169–87. doi:10.1089/ten.teb.2009.0352.

Wang C-H, Hsieh D-J, Periasamy S, Chuang C-T, Tseng F-W, Kuo J-C, Tarng Y-W. Regenerative porcine dermal collagen matrix developed by supercritical carbon dioxide extraction technology: Role in accelerated wound healing. *Materialia* (2020) 9:100576. doi:10.1016/j.mtla.2019.100576.

Wang E, Han J, Zhang X, Wu Y, Deng X-L. Efficacy of a mineralized collagen bone-grafting material for peri-implant bone defect reconstruction in mini pigs. *Regen Biomater* (2019) 6:107–11. doi:10.1093/rb/rby029.

Wang J, Zhang Y, Aghda NH, Pillai AR, Thakkar R, Nokhodchi A, Maniruzzaman M. Emerging 3D printing technologies for drug delivery devices: Current status and future perspective. *Adv Drug Deliv Rev* (2021) 174:294–316. doi:10.1016/j.addr.2021.04.019.

Wang L, Wang N, Zhang W, Cheng X, Yan Z, Shao G, Wang X, Wang R, Fu C. Therapeutic peptides: current applications and future directions. *Signal Transduct Target Ther* (2022) 7:48. doi:10.1038/s41392-022-00904-4.

Wang Z, Li M, Wang B, Xu Y, Li J, Zhang S, Qin Q, Wang J. Biomimetic niche of vascular intima with biophysical orientation and biochemical stimulation for rapid endothelialization and long-term patency. *Chem Eng J* (2023) 451:138805. doi:10.1016/j.cej.2022.138805.

Wang Z, Wang L, Li T, Liu S, Guo B, Huang W, Wu Y. 3D bioprinting in cardiac tissue engineering. *Theranostics* (2021) 11:7948–69. doi:10.7150/thno.61621.

Wang Zeyi, Liu Y, Wang Zhiwei, Huang X, Huang W. Hydrogel-based composites: Unlimited platforms for biosensors and diagnostics. *VIEW* (2021) 2:20200165. doi:10.1002/VIW.20200165.

Wenjin W, Wenchao L, Hao Z, Feng L, Yan W, Wodong S, Xianqun F, Wenlong D. Electrical stimulation promotes BDNF expression in spinal cord neurons through

Ca^{2+}- and erk-dependent signaling pathways. *Cell Mol Neurobiol* (2011) 31:459–67. doi:10.1007/s10571-010-9639-0.

Widenfalk J, Lipson A, Jubran M, Hofstetter C, Ebendal T, Cao Y, Olson L. Vascular endothelial growth factor improves functional outcome and decreases secondary degeneration in experimental spinal cord contusion injury. *Neuroscience* (2003) 120:951–60. doi:10.1016/S0306-4522(03)00399-3.

Williams SK, Touroo JS, Church KH, Hoying JB. Encapsulation of adipose stromal vascular fraction cells in alginate hydrogel spheroids using a direct-write three-dimensional printing system. *Biores Open Access* (2013) 2:448–54. doi:10.1089/biores.2013.0046.

Xian X, Gopal S, Couchman JR. Syndecans as receptors and organizers of the extracellular matrix. *Cell Tissue Res* (2010) 339:31–46. doi:10.1007/s00441-009-0829-3.

Xu T, Zhao W, Zhu J-M, Albanna MZ, Yoo JJ, Atala A. Complex heterogeneous tissue constructs containing multiple cell types prepared by inkjet printing technology. *Biomaterials* (2013) 34:130–9. doi:10.1016/j.biomaterials.2012.09.035.

Yan L, Zhou T, Ni R, Jia Z, Jiang Y, Guo T, Wang K, Chen X, Han L, Lu X. Adhesive gelatin-catechol complex reinforced poly(acrylic acid) hydrogel with enhanced toughness and cell affinity for cartilage regeneration. *ACS Appl Bio Mater* (2022) 5:4366–77. doi:10.1021/acsabm.2c00533.

Yanagawa F, Sugiura S, Kanamori T. Hydrogel microfabrication technology toward three dimensional tissue engineering. *Regen Ther* (2016) 3:45–57. doi:10.1016/j.reth.2016.02.007.

Yang J, Yang K, Man W, Zheng J, Cao Z, Yang C-Y, Kim K, Yang S, Hou Z, Wang G, Wang X. 3D bio-printed living nerve-like fibers refine the ecological niche for long-distance spinal cord injury regeneration. *Bioact Mater* (2023) 25:160–75. doi:10.1016/j.bioactmat.2023.01.023.

Yang K, Yang J, Man W, Meng Z, Yang C-Y, Cao Z, Liu J, Kim K, Liu Y, Yang S, Gua Y, He Z, Ma C, Wang G, Wang X. N-Cadherin-functionalized nanofiber hydrogel facilitates spinal cord injury repair by building a favorable niche for neural stem cells. *Adv Fiber Mater* (2023) 5:1349–66. doi:10.1007/s42765-023-00272-w.

Yoo J, Park C, Yi G, Lee D, Koo H. Active targeting strategies using biological ligands for nanoparticle drug delivery systems. *Cancers (Basel)* (2019) 11:640. doi:10.3390/cancers11050640.

Yurchenco PD, Amenta PS, Patton BL. Basement membrane assembly, stability and activities observed through a developmental lens. *Matrix Biol* (2004) 22:521–38. doi:10.1016/j.matbio.2003.10.006.

Zhang Y, Yu Y, Ozbolat IT. Direct bioprinting of vessel-like tubular microfluidic channels. *J Nanotechnol Eng Med* (2013) 4. doi:10.1115/1.4024398.

Zhang Z, Wang B, Hui D, Qiu J, Wang S. 3D bioprinting of soft materials-based regenerative vascular structures and tissues. *Compos Part B Eng* (2017) 123:279–91. doi:10.1016/j.compositesb.2017.05.011.

Zhao X, Song W, Chen Y, Liu S, Ren L. Collagen-based materials combined with microRNA for repairing cornea wounds and inhibiting scar formation. *Biomater Sci* (2019) 7:51–62. doi:10.1039/C8BM01054D.

Zhu K, Shin SR, van Kempen T, Li Y, Ponraj V, Nasajpour A, Mandla S, Hu N, Liu X, Leijten J, Lin Y-D, Hussain MA, Zhang, YS, Tamayol A, Khademshosseini A. Gold nanocomposite bioink for printing 3D cardiac constructs. *Adv Funct Mater* (2017) 27:1605352. doi:10.1002/adfm.201605352.

Chapter 2

Renal Proximal Tubule Injury: Role of Angiotensin II Type 1 Receptor (AT_1R) Blockage through Heat Shock Protein 70 (Hsp70)

Patricia G. Vallés[1,2,*]
Andrea Fernanda Gil Lorenzo[1]
Valeria Cacciamani[2]
María Eugenia Benardon[1]
and Valeria Costantino[2,3]
[1]Área de Fisiopatología, Departamento de Patología, Facultad de Ciencias Médicas, Universidad Nacional de Cuyo, Mendoza, Argentina
[2]IMBECU-CONICET (Instituto de Medicina y Biología Experimental de Cuyo – Consejo Nacional de Investigaciones Científicas y Técnicas)
[3]Área de Biología Celular, Departamento de Morfofisiología, Facultad de Ciencias Médicas, Universidad Nacional de Cuyo, Mendoza, Argentina

Abstract

The renal tubules and tubulointerstitial space make up a significant portion of the kidney and are the major sites in response to injuries. The severity of tubular interstitial damage is highly correlated to the degree of decline of renal function, even better than the glomerular lesions do. Increasing evidence shows that proximal tubular epithelial cells (PTECs) play diverse roles in renal repair or progression to chronic kidney disease (CKD). The innate immune characteristics of PTECs enable them to act

* Corresponding Author's E-mail: pvalles@fcm.uncu.edu.ar.

In: Advances in Biology. Volume 5
Editor: Charles D. Grant
ISBN: 979-8-89113-441-6

as immune responders to a wide range of insults, with the consequent production and release of bioactive mediators that drive interstitial inflammation and fibrosis.

The renin–angiotensin system is a critical promoter of fibrogenesis. It represents a nexus among glomerular capillary hypertension, barrier dysfunction and renal tubular injury caused by abnormally filtered proteins. Transforming growth factor (TGF-β) and reactive oxygen species (ROS) are the key mediators of the pro-fibrotic effect of angiotensin II causing apoptosis and epithelial-to-mesenchymal transition of the renal tubular epithelium. Angiotensin II via angiotensin type 1 receptor (AT_1R) regulates the activation and trans-differentiation of pericytes and fibrocytes into scar-forming myofibroblasts. Detachment and phenotypic changes of the former, can lead to the loss of peritubular capillaries and also contribute to hypoxia-dependent fibrosis. Many effects of angiotensin II are dependent on the AT_1R stimulation of reactive oxygen species (ROS) production by NADPH oxidase. Angiotensin II upregulation stimulates oxidative stress in proximal tubules. The NADPH oxidase 4 (Nox4) is abundantly expressed in kidney proximal tubule cells.

Induction of the stress response includes synthesis of heat shock protein 70, a molecular chaperone that has a critical role in the recovery of cells from stress and in renoprotection, guarding cells from subsequent insults. Hsp70 chaperone functions in part by driving the molecular triage decision, which determines whether proteins enter the productive folding pathway or result in client substrate ubiquitination and proteasomal degradation.

This review discusses the current mechanisms of tubulointerstitial fibrosis progression and the increasingly important role of the proximal tubule injury in driving inflammation and fibrosis. The renoprotective role of the AT_1R blockage through Hsp70, as a possible molecular target for therapeutic strategy against renal PTECs injury, is discussed.

Introduction

Chronic kidney disease (CKD) is a growing epidemic resulting from renal insults such as diabetes mellitus, hypertension, inflammatory glomerular diseases, genetic disorders and toxins, with a consistent estimated global prevalence of between 11 to 13% (N. R. Hill et al. 2016).

All stages of CKD are associated with increased risks of cardiovascular morbidity, premature mortality and/or decreased quality of life. CKD is usually asymptomatic until later stages and accurate prevalence data are lacking. Although some renal injuries may target the glomeruli versus the tubular compartment, the best predictor of renal survival in CKD of all etiologies is the amount of tubule interstitial fibrosis on kidney biopsy (Nath 1992; Risdon, Sloper, and De Wardener 1968). Accumulating evidence shows that renal function decline correlates better with tubulointerstitial damage than that of glomerular injury (Risdon, Sloper, and De Wardener 1968; Bohle et al. 1979). Maladaptive repair of injured tubules after acute kidney injury (AKI) also leads to the progression of CKD (Mackensen-Haen et al. 1981; Ferenbach and Bonventre 2015). Thus, proximal tubule epithelial cells (PTECs) should be regarded not only as victims in the context of kidney diseases, but also as key inflammatory and fibrogenic cells that drive the progression from acute to chronic kidney disease (Venkatachalam et al. 2015; B. C. Liu et al. 2018).

The combined effects of interstitial inflammation, oxidative stress and local angiotensin II activity result in the disruption of glomerulus-tubule continuity, the development of pathogenic hypoxia, the generation of myofibroblasts and fibrosis, and the impairment of the protective autoregulation of glomerular blood flow that leads to glomerulosclerosis (Rodríguez-Iturbe and García García 2010).

Tubulointerstitial inflammation can be observed in the early stage of many renal diseases, and the response of PTECs to injury is likely to be a key determinant in the development of interstitial inflammation (B. C. Liu et al. 2018).

Angiotensin II, the main effector of the renin–angiotensin system, is a critical promoter of fibrogenesis. It represents a nexus among glomerular capillary hypertension, barrier dysfunction, and renal tubular injury caused by abnormally filtered proteins. Transforming growth factor (TGF-β) and reactive oxygen species (ROS) are the key mediators of the pro-fibrotic effect of angiotensin II causing apoptosis and epithelial-to-mesenchymal transition (EMT) of the renal tubular epithelium (Macconi, Remuzzi, and Benigni 2014).

The production of ROS is facilitated by NADPH oxidase (Gill and Wilcox 2006). Many of the effects of the angiotensin II/AT_1R are dependent on ROS production stimulated by the activation of NADPH oxidase. ROS are important mediators of angiotensin II-induced epithelial mesenchymal transition (EMT) and apoptosis leading to renal fibrosis. Studies have suggested that caveolin 1 plays an important role as a platform for the compartmentalization of redox signaling events through NADPH oxidase–dependent production of ROS (Seikaly, Arant, and Seney 1990). Thus, it can be speculated that caveolin is involved in angiotensin II–induced NADPH oxidase activation in the kidney (Ushio-Fukai and Alexander 2006).

Seven NADPH oxidase isoforms have been identified (Nox1–5, Duox1, Duox2). The isoform Nox4, cloned from the kidney, is highly expressed in renal tubules, renal fibroblasts, glomerular mesangial cells, and podocytes (Schlecht et al. 2011).

A cell protective mechanism against oxidative stress includes the production of highly conserved proteins called heat shock proteins (Hsps). Hsps are intracellular proteins that respond to cellular stress and act as chaperones of other client' peptides to which they bind to prevent their irreversible aggregation and promote their correct folding, physiological assembly and intracellular trafficking (Gomez-Pastor, Burchfiel, and Thiele 2018). Hsp70 chaperones regulate a diverse set of signaling pathways through their interaction with proteins (Lorenzo et al. 2015).

This review focuses on the current mechanisms of tubulointerstitial fibrosis (TIF) progression and the growing important role of the proximal tubule injury in driving inflammation and fibrosis. The reno-protective role of the AT_1R blockage through Hsp70, as a possible molecular target for therapeutic strategy against renal PTECs injury, is discussed.

Proximal Tubules as a Target of Renal Injury

Renal tubulointerstitial fibrosis (TIF) progression involves many different cell types and growth factors that mediate crosstalk between these different cellular compartments. The proximal tubule has long been regarded as a target of injury. However, advancing evidence suggests that it also plays a prominent role in the development and progression of TIF (Gewin 2018).

The proximal tubule is a main segment involved in AKI because, in part, to its vulnerability to changes in oxygen delivery. Cells from proximal tubule are rich on mitochondria to support the energy necessary for this segment to reabsorb over 60% of the filtered water and electrolytes (Weinberg and Molitoris 2009). To facilitate this, the apical surface has a brush border which increases the surface area available for reabsorption. In addition, the innermost part of the proximal tubule, the S3 segment, is supplied by a unique countercurrent vasculature with relatively lower oxygen tensions. Thus, a mismatch of oxygen supply and demand occurs if there is any disruption to the blood supply or drop in blood pressure. If AKI is severe enough or repetitive, it leads to TIF and CKD. In addition, the proximal tubule has multiple transporters that render this segment susceptible to chronic injury from toxins, drugs or filtered proteins.

The nature of injury (severity and duration) and the proximal tubular response to chronic injury dictate whether repair or TIF progression occurs. Some of these proximal tubular responses to injury are initially adaptive but become detrimental in the context of CKD.

These maladaptive responses can promote tubular atrophy and TIF both through cell autonomous effects that alter cell survival or extracellular matrix (ECM) production and by paracrine effects on surrounding cells (Schnaper 2017).

Proximal Tubule Epithelial Cells (PTECs) as Inflammatory Cells

PTECs to injury are likely to be a key determinant in the development of interstitial inflammation. The PTECs themself may have immunologic properties when appropriately activated, including phagocytosis and subsequent antigen presentation (V. R. Kelley and Singer 1993) as well as co-stimulation of dendritic cells (Macconi et al. 2009; Weisheit, Engel, and Kurts 2015) or lymphocytes (Y. Chen et al. 2006). The proximal tubular epithelial cell is activated to migrate and proliferate in order to replenish the tubular structure, but it also produces chemoattractants and fibrogenic factors.

These damaged PTECs can be transformed into a secretory phenotype and elicit proinflammatory mediators as proinflammatory cytokines (e.g., interleukin, tumor necrosis factor, colony-stimulating factor and growth factor) (B. C. Liu et al. 2018). Through these secreted factors, the injured proximal tubule has paracrine effects on surrounding compartments such as the vasculature, inflammatory system and interstitial fibroblasts.

Since the first report of tumor necrosis factor alfa (TNFα) and interleukin six (IL-6) produced by PTECs (Jevnikar et al. 1991; Yard et al. 1992), a variety of cytokines have been shown to be produced by activated PTECs, including IL-1b (Leemans et al. 2014; Anders 2016), IL18 (Yunbo Yang et al. 2015), IL-15 (Wong et al. 2003; Weiler et al. 2001), IL-16 (Wang S. et al. 2008), TNF-a, (Al-Lamki and Mayadas 2015), TWEAK (Sanz Ana B et al. 2014; Sanz Ana B., Sanchez-Nĩo, and Ortiz 2011), Fas ligand (Ortiz, Lorz, and Egido 1999; Lorz et al. 2000), connective tissue growth factor (CTGF) (Rodrigues-Diez et al. 2015; Sánchez-López et al. 2009) and vascular endothelial growth factor (Kang et al., 2001; Ninichuk et al., 2006).

Chemokines are a family of small molecular cytokines with chemotactic activity. PTECs are rich sources of the CC chemokine ligand (CCL) subfamily, including monocyte chemoattractant protein-1/CCL2, RANTES (regulated on activation, T-cell expressed, and secreted)/CCL5, monocyte chemoattractant protein-1/CCL3, and CX3CL subfamily (fractalkine/CX3CL1) (B. C. Liu et al. 2018; Chung and Lan 2011).

Monocyte chemoattractant protein-1/CCL2 one of the most widely studied chemokines in kidney injury (Yiping Wang et al. 1997; 2000; Furuichi et al. 2003). Increased levels of monocyte chemoattractant protein-1/CCL2 are associated with progressive tubulointerstitial inflammation (Segerer and Alpers 2003; Tesch 2008). Stimulation of all three MAPK pathways appears to be required for MCP-1 and IL-8 production (H. Li and Nord 2002).

Cell adhesion molecules such as the cadherins, catenins, zonula occludens protein 1, occludin, and claudins are essential for maintaining the epithelial polarity and barrier integrity that are necessary for the normal absorption/excretion of fluid and solutes. The CD40/CD154 ligation can induce mononuclear cell adhesion to PTECs via an intracellular adhesion molecule-1-dependent mechanism (H. Li and Nord 2005).

In relation to adhesion molecules, it is now widely appreciated that intracellular adhesion molecule-1 and selectins derived from PTECs represent relatively early events in the pathophysiology of renal injury by promoting leukocyte infiltration and inflammatory responses (H. Li and Nord 2005; Prozialeck and Edwards 2007).

Inflammation that does not resolve after a chronic injury can create a vicious cycle of inflammation, tissue damage and fibrosis.

Potential Mechanisms of Tubular Injury Driving the Progression of CKD

Participation of Innate Immune-Sensing Receptors in Proximal Tubule Epithelial Cells

Excessive or uncontrolled inflammatory responses can lead to progressive kidney injury. Taking into account the immune characteristics of PTECs, Toll-like receptors (TLRs) and NLRP3 inflammasome have important roles in the pathogenesis of multiple renal disorders.

First of all, tubular cell death initiates an innate immune response. Damaged or dying cells release endogenous molecules called damage/

danger-associated molecular patterns (DAMPs) system in a fashion analogous to pathogen associated molecular patterns (PAMPs), molecules released by pathogenic bacteria or viruses. Both molecule patterns, DAMPs and PAMPs activate identical pattern recognition receptors such as Toll-like receptors on tissue-resident cells (dendritic cells, fibroblasts, and tubular cells) or recruited leukocytes (Kurts et al. 2013; Wu et al. 2007) causing them to secrete proinflammatory cytokines and chemokines (Anders and Schaefer 2014; Kurts et al. 2013; D. Yang, Han, and Oppenheim 2017).

TLRs are a family of transmembrane receptors, and the signal transduction initiated from TLRs activates effector cells via several kinases and NF-kB dependent mechanisms, associated to renal damage (Gluba et al. 2010). TLRs are widely expressed in TECs. PTECs are known to express both TLR2 and TLR4, and their signaling are activated during ischemia/ reperfusion injury (Mkaddem et al. 2009; Wu et al. 2010; Allam et al. 2012; Wolfs et al. 2002) and sepsis-induced AKI (El-Achkar and Dagher 2006; El-Achkar et al. 2006; Dear et al. 2006). A wide range of danger-associated molecular patterns (DAMPs) such as high-mobility group box1 protein, histones, and heat- shock proteins released by necrotic tubular cells can activate TLR2 and TLR4 on PTECs (Wu et al. 2010; Allam et al. 2012). Ding LH et al., also demonstrated that albumin might serve as an endogenous DAMP to trigger the activation of the TLR2-MyD88-NF-kB pathway (L. H. Ding et al. 2015).

In addition to PAMPS and DAMPS that bind to TLRs on the cell surface, the cytoplasmic NOD-like receptor protein 3 (NLRP3) inflammasome is integrated to the inflammatory cascade in sepsis- associated AKI (Roh and Sohn 2018; Gonçalves, Zamboni, and Cĝmara 2010).

NLRP3 inflammasome activation triggers the secretion of IL-1b and IL-18 via caspase-1 mediated processing of pro-IL-1b and pro-IL-18 and pyroptosis accounting for endotoxin-related pathology (Zamyatina and Heine 2020).

Liu et al. recently found that proteinuria causes NLRP3 inflammasome activation and IL-1b/IL-18 maturation in a time course and dose-dependent manner in the proximal tubule cells (D. Liu et al. 2014). Further investigation indicated that megalin/cubilin–mediated albumin retention and lysosomal rupture are involved in the activation of NLRP3 inflammasome and interstitial inflammation (D. Liu et al. 2015). Moreover, angiotensin II has also been shown to induce NLRP3 inflammasome activation in PTECs, which is associated with mitochondrial dysfunction or endoplasmic reticulum stress (J.

Wang et al. 2015; Wen et al. 2016). Thus, activation of the inflammasome pathway may represent a new mechanism of tubulointerstitial inflammation.

TLRs and inflammasomes result in the production of proinflammatory cytokines and chemokines by PTECs, the recruitment of inflammatory cells infiltrating the interstitial space, and the development of interstitial inflammation and fibrosis.

Proximal tubule (PT) injury can promote the occurrence of inflammation, but accumulating evidence also reveals the inhibitory effect of PT on inflammation, suggesting that PT may play a dual role in the occurrence and development of inflammation.

Human PTECs are able to and regulate antibody production in autologous B cells (Sampangi et al. 2015). Colony-stimulating factor-1 (CSF-1), produced primarily by PTECs, can stimulate resident renal macrophages and dendritic cells, participating in recovery from AKI in response to either injury/reperfusion or a model of selective PT injury induced by diphtheria-toxin-induced apoptosis in transgenic mice expressing the human diphtheria toxin receptor on PTECs (Yinqiu Wang et al. 2015). CSF-1 may be responsible for the polarization of renal macrophages and recovery from AKI.

Peroxisome proliferator-activated receptors (PPARs) are members of the nuclear hormone receptor superfamily of ligand-activated transcription factors. Accumulating evidence suggests that PPARs may play an important role in the pathogenesis of kidney disease.

All three members of the PPAR subfamily, PPARα, PPARβ/δ, and PPARg, have been implicated in many renal pathophysiological conditions, including AKI, diabetic nephropathy, and chronic kidney disease, among others. Preservation of proximal tubule peroxisome proliferator-activated receptor (PPARα) expression significantly reduces the expression of pro-inflammatory cytokines such as IL-1β, IL-6 and TNF-α, and significantly increases the expression of anti-inflammatory cytokines IL-10 and arginase-1, thereby reducing TIF and inflammation (S. Li et al. 2013). Emerging data suggest that PPAR agonists may be potential therapeutic targets for renal disease.

In conclusion, although studies have indicated that proximal tubule cells have both pro-inflammatory and anti-inflammatory effects, it is needed to determine which effect predominates in the TIF process, and it is also necessary to investigate whether the pro-inflammatory and anti-inflammatory mechanisms of PT interact or not.

Myofibroblast: The Key Extracellular Matrix Producer. Contribution of Renal PTECs to the Development of TIF

The inflammatory microenvironment that develops after renal injury activates myofibroblasts to produce matrix, which is a central event in TIF. Myofibroblasts are mesenchymal cells with increased stress fiber formation allowing for contractile properties and increased matrix production. Myofibroblast formation, proliferation and ECM production are regulated by various growth factors and hormones (Lebleu et al. 2013). Clarifying the origin of myofibroblasts is particularly crucial because they are producer of ECM.

It has been commonly assumed that fibroblasts are the primary source of myofibroblasts (Szeto et al. 2016; Machida et al. 2010). Most of the data implicate resident fibroblasts and/ or pericytes as the likely precursor to the myofibroblast, but some studies suggest that inflammatory cells and bone-marrow derived "fibrocytes" may become matrix-producing interstitial cells (Lebleu et al. 2013; Boor, Ostendorf, and Floege 2010; Mack 2018).

Angiotensin II via AT_1R regulates the activation and trans-differentiation of pericytes and fibrocytes into scar-forming myofibroblasts. Detachment and phenotypic changes of the former can lead to the loss of peritubular capillaries and also contribute to hypoxia-dependent fibrosis.

It has been shown that PTECs, endothelial cells, podocytes, pericytes, mesangial cells and macrophages are capable of converting into myofibroblasts and producing ECM (Machida et al. 2010; B. C. Liu, Tang, and Lv 2019; Y. B. Y. Sun et al. 2016).

Since renal tubules make up the largest proportion of the renal parenchyma, the process of epithelial-mesenchymal transition (EMT) cannot be ignored. EMT, a cellular phenotypic transformation process that occurs during embryonic development, tumor metastasis and organ fibrosis, has received a surge of attention in the development of TIF in recent years (Nieto et al. 2016; Sleeman and Thiery 2011). A large proportion of myofibroblasts are transformed from PTECs by EMT (Iwano et al. 2002).

The injured proximal tubule cells produce growth factors such as TGF-β, Wnt, and PDGF-β that stimulate the myofibroblast formation, proliferation and production of ECM (Koesters et al. 2010; Y. T. Chen et al. 2011; Maarouf et al. 2016; Iwano et al. 2002; F. T. Borges et al. 2013).

Due to the increased proliferation and decreased apoptosis of resident interstitial cells, there is an increase in interstitial fibroblasts at the tubulointerstitial space. Behind proliferation, these cells may undergo a phenotypic transformation to fibrosis-promoting properties. This phenotypic change can be recognized by the acquisition of smooth muscle cell-type properties and *de novo* expression of α smooth muscle actin (α-SMA). α-SMA-expressing cells, myofibroblasts, are restricted to periarteriolar regions in normal kidneys and they densely populate the interstitial space of chronically damaged kidneys. Compressed by collagen, peritubular capillaries become atrophic and distance between tubular cells and capillaries increases, thereby worsening the damage and facilitating the development of renal failure (Freedman and Cohen 2016).

Loss of the microvasculature implies a hypoxic milieu and suggests an important role for hypoxia. Microvasculature dysfunction, by inducing a hypoxic environment, triggers inflammation, EMT with myofibroblast differentiation and fibrosis (G. S. Hill 2008; Fine and Norman 2008; S. Sun et al. 2009). In vitro studies have largely focused on PTECs as the predominant epithelial cell type in the cortex and susceptible to hypoxic injury.

Exposure of PTECs to hypoxia induces a myofibroblastic phenotype whereas more prolonged exposure leads to mitochondrial injury and apoptosis consistent with the loss of tubular cells *in vivo* (Nangaku 2006; Nangaku and Eckardt 2007). Data from animal models provides a strong argument for hypoxia as a primary mediator of progressive scarring in the kidney, however a critical question is whether this also applies to hypertension in humans (Heyman et al. 2008). Although, there are currently only limited clinical data, the relationship between rarefaction of peritubular capillaries and tubulointerstitial scarring in human biopsies strongly suggests a role of hypoxia in chronic kidney disease.

Involvement of Microvascular Injury on Tubulointerstitial Fibrosis (TIF)

Hypoxia may prevent renal tubular recovery and re-differentiation after injury, as suggested by the strong association between peritubular capillary rarefaction and tubular de-differentiation (Polichnowski et al. 2014).

Biopsies of patients with CKD show significant capillary rarefaction concurrent with TIF. Sustained renal injury is accompanied by a process of de-differentiation, and the reduced response of de-differentiated cells to decreased adenosine triphosphate (ATP) and energy supply allows the cells to survive under hypoxic conditions (Kang et al. 2015). When proximal tubule is damaged, the PTEC de-differentiates and loses expression of typical epithelial proteins such as E-Cadherin and ZO-1. Acutely, this may be beneficial as a de-differentiated cell should require less ATP and energy to maintain a benefit in hypoxic conditions.

However, de-differentiation cannot be simply defined as a beneficial process, as it has been shown in unilateral ureteral obstruction and nephrotoxic serum–induced nephritis, nephropathy models that TIF is aggravated when genetically is targeting Snail1 expression, a key transcription factor in PTECs de-differentiation (Grande et al. 2015).

The mechanism of sustained epithelial cell de-differentiation may, on the one hand, be through the production of paracrine factors, such as TGF-β, to stimulate the production of ECM by adjacent pericytes/myofibroblasts to induce TIF progression and on the other hand, it may promote TIF by exacerbating ECM deposition through cell-autonomous production of collagen proteins such as COL1A2 protein (Fragiadaki et al. 2011; Okada et al. 2005).

One unanswered question regarding this spectrum of de-differentiation is, how it relates functionally to adaptive or maladaptive repair.

Cell-Cycle Arrest

Most proximal tubule epithelia in the uninjured kidney are quiescent (cell cycle stage G0). However, during injury, cells may enter the cell cycle (G1, S, G2, M) to help replace cells lost to apoptosis/necrosis, and some cells become arrested in either G1 or G2. This cell arrest is adaptive to allow time for repair of any DNA damage and prevent the propagation of mutations that occur in injured cells. However, persistence of G2 arrest in epithelial cells leads to increased production of profibrotic cytokines TGF-β and CTGF in a JNK-dependent pathway (L. Yang et al. 2010; Lovisa et al. 2015).

Expression and Synthesis of the Components of the Renin-Angiotensin System (RAS) in Renal Proximal Tubules

The renin–angiotensin system is a critical promoter of fibrogenesis. It represents a nexus among glomerular capillary hypertension, barrier dysfunction and renal tubular injury caused by abnormally filtered proteins. Transforming growth factor beta (TGF-β) and reactive oxygen species (ROS) are the key mediators of the profibrotic effect of angiotensin II causing apoptosis and EMT of the renal tubular epithelium. The activity of local RAS is essential for the mechanisms mediating pathophysiological functions of renal injury (Navar et al. 2002). Renal proximal tubules have been recognized as a major target of both circulating and local paracrine angiotensin II under physiological conditions as well as in hypertension (Harris and Navar 1985; Zhuo and Li 2013; Chappell 2012). Indeed, all major components of the classical RAS, including the sole substrate angiotensinogen (AGT), two key enzymes renin and angiotensin-converting enzyme (ACE), and two major G protein–coupled receptor (GPCR) subtypes of AT_1 and AT_2, are localized in the proximal tubules of the kidney (Mendelsohn 1982). The expression of AGT, renin and ACE in the proximal tubules is crucial for the onsite production of intratubular and intracellular angiotensin II (Gabriel Navar and Nishiyama 2001). The major fraction of angiotensin II present in renal tissues is generated locally from angiotensinogen delivered to the kidney as well as from angiotensinogen locally produced by proximal tubule cells.

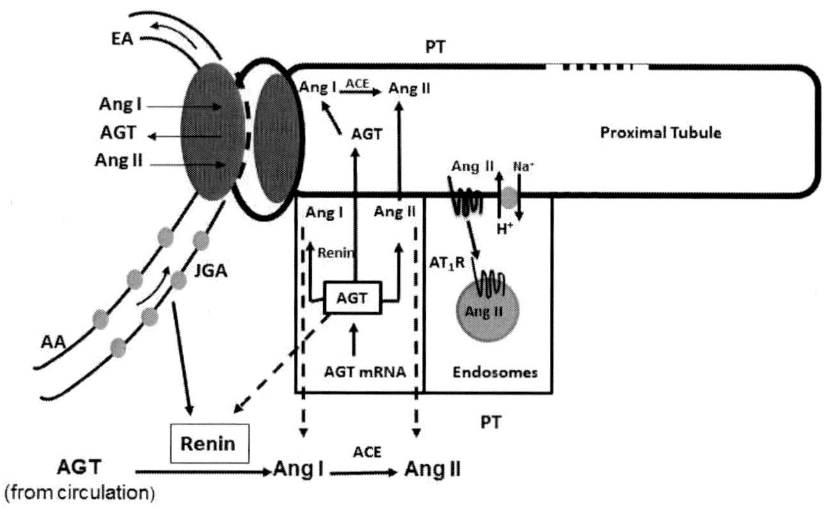

Figure 1. Intrarenal renin-angiotensin system (RAS).
The expression of AGT, renin, and ACE in the proximal tubules is crucial for the onsite production of intratubular and intracellular angiotensin II. Angiotensin II is also internalized into endosomes via the AT_1 receptor. AA, afferent arteriole; EA, efferent arteriole; JGA, juxtaglomerular apparatus; AGT, angiotensinogen; Ang, angiotensin, ACE, angiotensin-converting enzyme; AI, angiotensin I; AII, angiotensin II; AT_1R, angiotensin II type 1 receptor; PTCs, proximal tubule cells. See text for details and references.

It has been further observed that double transgenic mice, expressing human renin systemically and human angiotensinogen in proximal tubular cells, show increased intrarenal angiotensin II levels with unchanged plasma angiotensin II levels (Kobori et al. 2007).

High levels of angiotensin II present in interstitial and intratubular fluid compartments combined with expression of abundant AT_1 receptors in both apical and basolateral membranes promote cellular uptake through receptor-mediated endocytosis with subsequent accumulation within the proximal tubule cells (X. C. Li and Zhuo 2014) (Figure 1). The levels of angiotensinogen, angiotensin I, and angiotensin II in the proximal tubular fluid often greatly exceed those in plasma and the whole kidney (Zhuo and Li 2011).

Intratubular angiotensin II could be derived from locally formed and secreted angiotensinogen (Figure 1). The angiotensinogen produced in proximal tubule cells seems to be secreted directly into the tubular lumen (Figure 1). Once angiotensin I is formed, conversion readily occurs because there are abundant amounts of ACE associated with the present in tubular fluid throughout the nephron being higher at the initial portion of the proximal tubule (Casarini et al. 1997). The localization of AT_1 and AT_2 receptors in the proximal tubules plays a key role in mediating biological and physiological effects of angiotensin II. AT_1 receptor mediated uptake of angiotensin II by the kidney, especially by the proximal tubules, plays a key role in intrarenal accumulation of the circulating angiotensin II (Zhuo et al. 2002; X. C. Li et al. 2007). Indeed, fluorescently labeled angiotensin II (FITC-Angiotensin II) was found to bind to the AT_1R followed by internalization to the endosomal and nuclear compartments in cultured rabbit and mouse proximal tubule cells (X. C. Li et al. 2006). Connected with this, caveolin 1 has been demonstrated to be required for normal renal AT_1 receptor expression (Ishizaka et al. 1998). This protein is necessary for correct transport of the AT_1R to the plasma membrane (Wyse et al. 2003). Treatment with angiotensin II results in the partial re-localization of activated AT_1 receptor into lipid rafts (cyclodextrin-sensitive cluster). Nonetheless, mislocalized caveolin mutant expression, absence of caveolin, or the disruption of the formation of a caveolin-AT_1 receptor complex have a profound effect on the trafficking of AT_1R to the plasma membrane. Moreover, caveolin 1 null mice have 55% reduction in renal AT_1 receptor levels compared with controls (Zhao et al. 2002). Together these data may suggest a chaperone role for caveolin in trafficking a transmembrane receptor to the cell surface (Wyse et al. 2003). Previously, we showed intensive decrease on caveolin 1 protein expression in microdissected proximal tubule membrane fraction from spontaneously hypertensive rats (SHR) without treatment compared with control, whereas higher caveolin 1 protein levels were detected in SHR after angiotensin II AT_1 receptor antagonist treatment. It is possible that in SHR proximal tubule epithelial cell membranes, angiotensin II stimulation results in internalization of caveolae and translocation to intracellular compartments in which the caveolin 1 protein is degraded (Bocanegra et al. 2010).

Angiotensin II Gives Rise to Proteinuria and Tubulointerstitial Fibrosis

Angiotensin II-induced proteinuria is a leading cause of CKD progression. After nephron loss, Brenner et al. identified glomerular capillary hypertension as a leading cause of progressive deterioration of remaining nephrons (Bohrer et al. 1977). Elevated transcapillary hydraulic pressure increased the population of large unselective pores perforating the glomerular membrane by a mechanism at least partly mediated by angiotensin II (Bohrer et al. 1977). Mechanical strain, as results of glomerular hypertension, upregulates AT_1R and increases the production of angiotensin II in cultured podocytes (Durvasula et al. 2004). Loss of podocyte–podocyte contact and increased albumin permeability induced by angiotensin II in these cells (Macconi et al. 2006) translate in the glomerular sieving dysfunction, observed in proteinuric nephropathies.

Angiotensin II constitutes a link among glomerular capillary hypertension, barrier dysfunction, and renal tubular injury caused by abnormally filtered proteins. TGF-β and reactive oxygen species (ROS) are the key mediators of the profibrotic effect of angiotensin II causing apoptosis and EMT of the renal tubular epithelium (Macconi, Remuzzi, and Benigni 2014).

The proteinuric ultrafiltrate represents the way for spreading the disease from the glomerulus to the tubulointerstitial compartment. Proteins abnormally filtered through the glomerular capillary have intrinsic toxicity on the proximal tubule contributing to the development of interstitial inflammation, fibrosis and ultimately to kidney dysfunction.

Proximal tubular cells exposed *in vitro* to albumin undergo apoptosis through a mechanism involving reduced expression of its receptor megalin. Furthermore, overload of plasma proteins stimulates proximal tubular cells to synthesize and release pro-inflammatory substances including monocyte chemoattractant protein-1 (MCP-1/CCL2), normal T-cell expressed and secreted (RANTES/CCL5), fractalkine/CX3CL1 that are potent chemoattractants for monocytes/macrophages and T lymphocytes.

Changes observed *in vitro* parallel those obtained in proteinuric models *in vivo*. They showed apoptosis-induced glomerular–tubule disconnection and inflammation via release of cytokine including MCP-1 from activated proximal tubular cells (Macconi, Remuzzi, and Benigni 2014). A similar inflammatory pathway occurs in patients with severe proteinuria in whom renal upregulation of MCP-1, RANTES, and osteopontin, mainly in tubular epithelial cells, correlates with the severity of the disease (Mezzano et al. 2000; 2001). Both injured tubules and inflammatory cells enhance fibrogenesis by activating interstitial fibroblasts via a paracrine mechanism that involves the release of TGF-β, a potent inducer of EMT.

Albumin induces the upregulation of TGF-β receptor type II expression in proximal tubular cells and makes them more susceptible to the matrix-stimulatory actions of TGF-β (Macconi, Remuzzi, and Benigni 2014).

Angiotensin II and TGF-β in Renal Fibrosis

Aside from protein load, angiotensin II directly stimulates the TGF-β gene, protein expression in proximal tubular cells (Weigert et al. 2002) and triggers the production of plasminogen activator inhibitor-1 (PAI-1), ECM synthesis and deposition in the interstitial space (Wolf et al. 1996; Fintha et al. 2007).

Several intracellular signaling pathways are activated by angiotensin II, to mediate renal fibrosis and inflammation, including TGF-β/Smads, nuclear factor-kappa B (NF-kB), and mitogen-activated protein kinases (MAPK). TGF-β signaling includes Smad and non-Smad pathways for the activation of its major effector and downstream target PAI-1 (Wei et al. 2013). Kindlin-2, a b-integrin adaptor protein has been recently discovered to work as a TGF-β type I receptor/Smad3 adapter protein that amplifies profibrogenic effects of TGF-β through Smad3 signaling in tubular cells *in vitro* and *in vivo* (Wei et al. 2013). Indeed, many genes involved in tissue fibrosis (e.g., ColIa1, ColIa2, ColIIIa1, ColVa2, ColVIa1, ColVIa3 and tissue inhibitor of matrix metalloproteinase-1 are regulated by TGF-β/Smad3 signaling (Verrecchia, Chu, and Mauviel 2001). Thus, Smad3 plays an essential role in angiotensin II- mediated fibrosis *in vivo* and *in vitro* (W. Wang et al. 2006; F. Yang et al. 2009; Meng et al. 2010; Z. Liu, Huang, and Lan 2012; Huang et al. 2010).

The non-Smad pathway involves reactive oxygen species (ROS)-dependent-c-Src-mediated activation of epidermal growth factor receptor and downstream signaling cascade (Kavsak et al. 2000). Control of ECM accumulation through the inhibition of TGF-β/PAI-1 has been proposed as the underlying mechanism of the therapeutic effect of angiotensin II blockade in several models of progressive nephropathies (Ebisawa et al. 2001).

Smad7 is an inhibitory Smad that negatively regulates TGF- β/ Smad-mediated renal fibrosis by facilitating degradation of TGF- β receptor-1 and Smads (Ebisawa et al. 2001; Tan et al. 2008; F. Y. Liu et al. 2007).

In chronic kidney diseases, many mediators such as TGF-β and angiotensin II are able to induce Smad7 mRNA expression, but Smad7 protein is degraded. In renal fibrosis, Smad7 protein levels are reduced. Since Smad7 functions as an integrated inhibitor for both TGF-β/Smad–signaling and NF-κB–signaling pathways, once renal Smad7 is degraded, TGF-β/Smad and NF-κB signaling becomes activated, resulting in the development of renal fibrosis and inflammation (G. X. Liu et al. 2013).

In addition, Li et al. have previously shown that Smad7 is renoprotective, as deletion of Smad7 enhances whereas overexpression of renal Smad7 inhibits angiotensin II–induced hypertensive heart and kidney injury (G. X. Liu et al. 2014). Consistent with this notion, it has been shown that a single deletion of ACE2 or Mas sustained AT_1-NF-κB-driven renal inflammation and fibrosis, which became more severe in mice lacking both ACE2/Mas genes. Thus, enhanced AT_1-ERK1/2-Smad3 signaling may be a mechanism by which loss of the ACE2/angiotensin 1–7/Mas axis exacerbated renal fibrosis in ACE2/Mas double KO mice. Furthermore, angiotensin II can also activate NF-κB by degrading Smad7 (Ni et al. 2020).

Previously, in the kidneys of angiotensin II infused-mice a significant upregulation of TGF-β and P-Smad3 has been demonstrated. MicroRNAs (miRNAs or miRs), small (about 22 nucleotides) noncoding RNAs regulate gene expression via inhibiting translation or guiding corresponding target mRNA cleavage. Ding et al. have shown that MiR-101a overexpression significantly suppressed the angiotensin II-induced increase of TGF-βR (one of serine/threonine kinase receptors of TGF-β/Smad signalling) and p-Smad3 and NF-κB. Hence, these data suggested that miR-101a may exhibit anti-fibrosis activities through inhibiting the TGF-β/Smad3 signaling pathway. Moreover, overexpression of miR-101a, inhibited angiotensin II-stimulated up-regulation of pro-inflammatory cytokines and decreased the phosphorylation level of IκBα and p65, which meant that miR-101a also acted as an inhibitor of the NF-κB signaling pathway in renal inflammation (H. Ding, Zhou, and Huang 2019).

Angiotensin II and Reactive Oxygen Species (ROS)

ROS are important mediators of Angiotensin II-induced EMT and apoptosis leading to renal fibrosis. Many effects of angiotensin II are dependent on the AT_1R stimulation that leads to reactive oxygen species (ROS) production by NADPH oxidase. From the seven NADPH oxidase isoforms that have been identified, the NADPH oxidase 4 (Nox4) is abundantly expressed in kidney PTECs. ROS generated by NADPH oxidase can also act as signaling molecules that mediate angiotensin II–dependent signal transduction pathways (redox-sensitive signaling) (Gill and Wilcox 2006; Hitomi, Kiyomoto, and Nishiyama 2007). Caveolin 1 plays an important role as a platform for the compartmentalization of redox signaling events through NADPH oxidase–dependent production of ROS (Seikaly, Arant, and Seney 1990).

ROS function as signaling molecules and contribute to cell growth, migration, differentiation, and cytoskeletal remodeling (Haugen, Croatt, and Nath 2000) being the major pathway by which angiotensin II induces epithelial–mesenchymal transition and apoptosis leading to fibrosis.

Previously, we showed that NADPH oxidase activity and Nox4 protein levels were increased in microdissected proximal tubule cell membranes from spontaneous hypertensive rats (SHR), an effect which was prevented by losartan treatment. Interestingly, losartan markedly increased proximal tubular caveolin 1, suggesting that intrarenal Angiotensin II is involved in the negative regulation of caveolin 1 during the development of hypertension and renal injury in SHR (Bocanegra et al. 2010).

It has been shown that treatment with a specific NADPH oxidase inhibitor, apocynin, decreased NADPH oxidase activity and attenuated PTECs apoptosis and tubulointerstitial fibrosis in mice with proximal tubular cell–specific overexpression of angiotensinogen. These data are consistent with the concept that locally generated angiotensin II in proximal tubules induces renal injury through NADPH oxidase-dependent ROS production (F. Liu et al. 2009).

ROS are involved in the activation of hypoxia-inducible factor-1 (HIF-1), a heterodimeric transcription factor that has a key role in cellular adaptation to different oxygen concentrations. Angiotensin II activates HIF-1a through the generation of ROS that increase HIF-1a translation and inactivate HIF prolyl-hydroxylases (PHD) promoting HIF-1a stabilization under non-hypoxic conditions (Patten et al. 2010).

A recent study has demonstrated a functional link between the activation of HIF-1a and the pro-fibrotic effects of angiotensin II in renal medullary interstitial cells (Zhengchao Wang et al. 2011). HIF-1α stimulates collagen accumulation by activation of fibrogenic factors, such as plasminogen activator inhibitor (PAI) and tissue inhibitor of metalloproteinase (TIMP-2). HIF prolyl-hydroxylases (PHD) are the enzymes that promote the degradation of HIF-1α. The mechanism whereby angiotensin II stabilizes HIF-1a in these cells is dependent on H_2O_2-mediated inhibition of PHD2 activity, the predominant PHD in the kidney. Gene silencing of HIF-1a blocks angiotensin

II-induced transcription of pro-fibrotic markers TIMP-1 and collagen I/III as well as of proliferating nuclear antigen and vimentin. *In vivo,* kidneys infused with angiotensin II show positive staining for both HIF-1a and a-smooth muscle actin in renal medullary interstitial cells (Zhengchao Wang et al. 2011). Altogether, these findings suggest that angiotensin II-induced interstitial fibrosis mediated by HIF-1a requires cell transdifferentiation possibly including EMT.

Interestingly, targeting angiotensin II signaling by angiotensin receptor blocker valsartan reduces renal expression of HIF-1a and its target gene TIMP-1 and attenuates interstitial fibrosis in streptozotocin-diabetic rats (Tang et al. 2012).

Zhou et al. have shown that angiotensin II activates nuclear transcription factor NF-κβ in proximal tubules, leading to long-lasting inflammatory and growth-promoting effects; these responses were blocked by inhibition of AT_1 receptor–mediated endocytosis of extracellular angiotensin II with losartan (Zhuo, Carretero, and Li 2006). The peptide angiotensin II is also involved in the inflammatory process through the synthesis of chemotactic factors, such as monocyte chemoattractant protein-1 (MCP-1) and by activation of resident cells by macrophage-related factors, and therefore contributes to the progression of fibrosis. Moreover, angiotensin II promotes the phenotypic change of fibroblasts to myofibroblasts (α-smooth muscle actin–positive cells). These activated fibroblasts may proliferate and invade the periglomerular and peritubular spaces, contributing to matrix deposition in the tubulointerstitial area (S. A. Mezzano, Ruiz-Ortega, and Egido 2001).

Heat Shock Protein 70 Function and Chaperone Cycle

Induction of the stress response includes synthesis of heat shock protein Hsp70, a molecular chaperone that has a critical role in the recovery of cells from stress and in cytoprotection, guarding cells from subsequent insults (Balchin, Hayer-Hartl, and Hartl 2016). The Hsp70 chaperones function in a wide range of cellular housekeeping activities, including the folding of newly synthesized proteins, the translocation of polypeptides into mitochondria and the endoplasmic reticulum (ER), the disassembly of protein complexes and the regulation of protein activity. Furthermore, Hsp70s prevent the aggregation and promote the refolding of misfolded denatured proteins, solubilize aggregated proteins and cooperate with cellular degradation machineries to clear aberrant proteins and protein aggregates (Figure 2). Thus, Hsp70s act as sentinel chaperones, guarding cells from the deleterious effects of a wide range of proteotoxic stresses (Balchin, Hayer-Hartl, and Hartl 2016; Nillegoda, Wentink, and Bukau 2018; Craig 2018; Mogk, Bukau, and Kampinga 2018; Fernández-Fernández et al. 2017) pathophysiological conditions (Qu et al. 2015) and organismal ageing (Walther et al. 2015) that cause protein homeostasis imbalance (Rosenzweig et al. 2019).

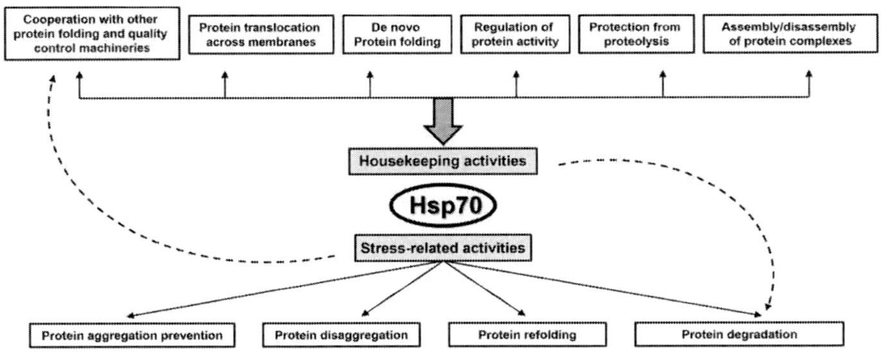

Figure 2. The diverse housekeeping and stress-related activities of Hsp70s.
The Hsp70 chaperones participate in a wide range of cellular housekeeping functions (top), including the folding of newly synthesized proteins, the translocation of polypeptides into mitochondria, chloroplasts and the endoplasmic reticulum (ER), the assembly and disassembly of protein complexes, regulation of protein activity and assisting in protein handover to downstream chaperone machineries, such as the Hsp90 folding machinery and chaperonins.
Furthermore, Hsp70s also participate in stress-related activities (bottom), such as preventing the aggregation of proteins, solubilizing aggregated proteins, promoting the refolding of misfolded or unfolded proteins and cooperating with cellular degradation machineries, such as autophagy and the ubiquitin–proteasome system (UPS), to clear aberrant proteins and protein aggregates.

Heat shock protein 70 (Hsp70) protects stressed cells by its ability to recognize nascent polypeptides, unstructured regions of proteins, and exposed hydrophobic stretches of amino acids (Kampinga et al. 2009).

The human Hsp70 family consists of 13 members, now systematically designated as the HSPA class (Kampinga et al. 2009). Prominent members are HSPA8 (HSC70, HSP73) which is constitutively expressed in all renal compartments (Komatsuda et al. 1992) and the major inducible forms HSPA1A (HSP72, HSP70i), HSPA1B (HSP70–2) and HSPA6.

All Hsp70 family members share at least two of the four structural features of the archetype Hsp70, the bacterial DnaK: an N-terminal, 45-kDa nucleotide binding domain (NBD), followed by a 15-kDa substrate binding domain (SBDβ), a 10-kDa helical lid domain (SBDα) and a disordered C-terminal tail of variable length.

In eukaryotic cytosolic and nuclear Hsp70s, the disordered tail ends frequently with a conserved charged motif (Glu-Glu-Val-Asp; EEVD) that interacts with specific cofactors (Zuiderweg, Hightower, and Gestwicki 2017). The NBD and SBDβ are connected by a conserved flexible linker that is essential for the allosteric mechanism that couples nucleotide binding and ATP hydrolysis to substrate binding (Vogel, Mayer, and Bukau 2006; Zhuravleva and Gierasch 2011; Alderson et al. 2014).

Hsp70 family members exhibit a high degree of specialization. Hsp70s can be distinguished by three main factors: first, differences in substrate recognition and allosteric regulation; second, adaptations imposed by target site localization and third, the type of cooperation with the extensive network of co-chaperones of the J-domain protein (JDP) family (also referred to as DnaJ, Hsp40 (164) and nucleotide exchange factors (NEFs) that control substrate presentation and release from Hsp70s (Abrams et al. 2014; Bracher and Verghese 2015).

In relation to Hsp70 allosteric cycle, the chaperone activity requires rapid association and timely release of substrates to prevent substrate aggregation and promote folding. This ability to associate with and dissociate from substrates in a controlled manner critically depends on an intricate allosteric mechanism that couples ATP hydrolysis in the NBD to substrate capturing by the SBD. ATP binds to the NBD, while misfolded client proteins interact with the SBD (Rosenzweig et al. 2019).

All Hsp70 chaperone activities require the regulatory and substrate targeting function of co-chaperones of the JDP family. JDPs are a heterogeneous group of modular multidomain proteins characterized by the J-domain, an α-helical hairpin domain approximately 70 amino acids long that is required to bind to Hsp70s and stimulate their ATPase activity (W. L. Kelley 1998). The association of JDPs with substrates initiates the entire Hsp70 chaperone cycle and therefore is of crucial importance.

Nucleotide exchange factors (NEFs) mediate the opening of the Hsp70 nucleotide binding cleft to facilitate the release of ADP, which in turn allows the rebinding of ATP and promotes the release of substrates from Hsp70 (Rosenzweig et al. 2019).

Accordingly, there are at least two conceptual ways of targeting Hsp70 family members: block their ATPase activity or change its protein-protein interactions with co-chaperones. Complexes between Hsp70 chaperones and co- chaperones have been shown to have specific functions, such as pro-folding, pro-degradation, and pro-trafficking. Hence, a promising strategy may be to block protein-protein interactions between HSP70s and their co-chaperones or to target allosteric sites that disrupt these contacts (A. Assimon et al. 2013).

Hsp70s intracellular locations are the cytosol, the nucleus and the mitochondria (O'Neill et al. 2014). Members of the Hsp70 family are called the "triage" chaperones because they play key roles in both protein folding and turnover (Chebotareva, Bobkova, and Shilov 2017). Hsp70 induction is an early survival response elaborated by stressed cells to counter cellular damage and hasten recovery. As central components of the cellular protein surveillance network, these ATP-dependent chaperones are involved in a large variety of protein folding processes. They promiscuously interact with practically all proteins in their unfolded, misfolded, or aggregated states but do not interact with their folded counterparts. In addition, a limited set of native proteins is recognized by Hsp70 and regulated in their activity, stability, and/or oligomeric state (Schlecht et al. 2011). Hsp70s interact with many different polypeptide conformers: with unfolded polypeptides, for example during de novo folding at the ribosome or translocation through biological membranes; partially structured on-path folding intermediates; protein species in off-path misfolded states; and native protein assemblies and non-native polypeptide aggregates. This high versatility of Hsp70 chaperones is due to the recognition of a short degenerate motif consisting of a core of five residues enriched in hydrophobic amino acids flanked by regions enriched in positively charged amino acids (Vogel, Mayer, and Bukau 2006). Such motifs occur on average every 30-40 residues in almost all proteins. In the native state, such motifs are mostly found in the hydrophobic core of the protein and are only exposed before de novo folding or upon denaturation, explaining the ability of Hsp70 chaperones to bind to misfolded proteins but not to their native counterpart (Mayer 2018).

After release from Hsp70s, the newly synthesized polypeptides can either fold spontaneously or be transferred for further folding and maturation to other chaperones, such as the Hsp90 system.

The co-chaperone Hop (known as Sti1 in yeast) facilitates substrate handover from Hsp70 to Hsp90 by bridging the two chaperones using its three tetratricopeptide repeat domains, as these domains bind the C-terminal EEVD motifs of the two chaperones (Scheufler et al. 2000; Odunuga et al. 2003).

The co-chaperone CHIP (Ballinger et al. 1999), an E3 ubiquitin-protein ligase, competes with Hop for binding to the C terminus of Hsp70 and Hsp90 (Stankiewicz et al. 2010) via its tetratricopeptide repeat domain. CHIP is a cytoplasmic protein whose amino acid sequence is highly conserved across species. The primary amino acid sequence is characterized by 3 domains, a tetratricopeptide repeat (TPR) domain at its amino terminus, a U-box domain at its carboxy terminus, and an interceding charged domain. The TPR domain and an adjacent charged region of CHIP (amino terminus residues 1 to 197) are necessary for its interaction with HSC70- HSP70i and CHIP interacts with the carboxy-terminal domain of HSC70 (residues 540 to 650 of HSC70), which is known to contain a TPR-acceptor site that also interacts with the TPR domain–containing co-chaperone HOP. Although the carboxy-terminal domain of the 70-kDa heat shock proteins is the interaction domain for CHIP's amino terminal TPR domain, it is the amino terminal ATP- binding domain of the HSP70 family members that regulates substrate binding in a nucleotide-dependent fashion (McDonough and Patterson 2003). CHIP functions as a link between Hsp70 chaperones and the proteasome systems, playing a vital role in maintaining the protein homeostasis in the cytoplasm.

Constitutive and Inducible Forms of Heat Shock Protein 70 in Kidney Tissue

HSC70 (70-kDa heat shock cognate protein), the major constitutively expressed member of the Hs70 family, can play an important role in the physiology and protection of the kidney tissue. In addition to being constitutively expressed, the synthesis of these proteins can be induced by a range of cellular insults (Welch 1993). The ubiquitous presence of HSC70 can be assumed to fulfill functions also in non- stressed cells, e.g., in assistance in protein folding, trafficking, and controlled protein degradation (Beck, Neuhofer, and Müller 2000), thus playing a role in protein maturation and cellular integrity. Similar levels of HSC70 have been observed in all zones of the normal rat kidney.

In the outer and inner medulla, staining for the chaperone is evident in proximal straight tubules, thick ascending limbs of the loop of Henle, collecting ducts, the papillary epithelium, and papillary interstitial cells

(Müller et al. 1996). Inducible Hsp70 chaperone (Hsp70i) is synthesized in response to environmental perturbations, such as ischemia, inflammation, and oxidative stress, as defending molecules that protect tissues from further injuries (Pockley 2003). Both ischemic and nephrotoxic renal failures induce Hsp70 in the cortex (Emami, Schwartz, and Borkan 1991) and outer medulla (Schober et al. 1997).

The protective properties of Hsp70i in AKI are not fully understood, and putative modes of protection include correction of protein conformation, cytoskeletal stabilization, anti-inflammatory effects, requirement in autophagy, anti-apoptotic properties, influence over macrophage phenotype, and stimulation of regulatory T cells (Tregs).

Hsp70i can interact with cytoskeletal structures and participate in the transport of proteins through intracellular membranes into the organelles and in the cleavage of protein aggregates. In the normal rat kidney, protein levels increase steeply along the corticopapillary axis (Müller et al. 1998) suggesting the protein involvement in the medullary cells' adaptation to hypertonic stress.

High extracellular NaCl concentration is a major determinant of the high Hsp70i expression. In the renal inner medulla, it has been suggested that during the initial phase of adaptation to high extracellular tonicities, when intracellular ionic strength is high due to cell shrinkage (Sone et al. 1993), this chaperone stabilizes intracellular proteins and thus reduces the denaturing effect of high intracellular concentrations of inorganic salts (Neuhofer et al. 2001).

In ischemic renal injury (Zhiyong Wang et al. 2011), it has been shown that Hsp70i promotes renal epithelial cell survival and preserves organ function by inhibiting proapoptotic changes in the activity of Akt kinase and glycogen synthase kinase 3β. Lentiviral-mediated Hsp70i repletion decreased mitochondrial Bax accumulation and rescued HSP70i knockout cells from death. In addition, the chaperone can limit pro-inflammatory NF-κB signaling in AKI by inhibiting NF-κB p65 translocation to the nucleus or marking proinflammatory Hsp90 client proteins for degradation (Aufricht et al. 1998).

Involvement of Hsp70i in the cytoskeletal stabilization was demonstrated in an animal model of postischemic AKI. During the early reflow period, basolateral membrane–resident proteins (Na^+-K^+-ATPase) were localized in the apical domain (Molitoris 1991; Van Why et al. 1992) reflecting the ischemia-induced loss of anchorage of Na^+-K^+-ATPase to cytoskeletal structures. The subsequent disappearance of Na^+-K^+-ATPase in the apical domain was accompanied by increased Na^+-K^+-ATPase in the membrane-cytoskeletal complex and parallels the accumulation of Hsp70i. ATP enhanced

Hsp70i release from the membrane-cytoskeletal complex and preserves Na^+-K^+-ATPase association with this complex (Aufricht et al. 1998). These findings are consistent with the idea that, during AKI, the inducible Hsp70 chaperone participates in the degradation of irreparably damaged proteins contributing to the restructuring of the cytoskeleton and restitution of epithelial cell polarity (Aufricht et al. 1998).

In the experimental model of low-protein (LP) diet, injury is induced from energy depletion in renal epithelial cells. Previously, we demonstrated that Hsp70i stabilizes renal Na^+/K^+-ATPase attachment to the cytoskeleton from the cortex and the outer stripe of the outer medulla during recovery from a LP diet. These results may suggest that the chaperone has a critical role in protecting the integrity of the cytoskeletal anchorage of Na^+/K^+-ATPase during recovery from ATP-depleted injury resulting from low- protein diet in the outer stripe of the outer medulla (Ruete, Carrizo, and Vallés 2008).

Upregulation of autophagy in renal tubular cells has been demonstrated to ameliorate both ischemia- reperfusion injury and cisplatin-induced AKI (Kaushal 2012). Proximal tubule-specific Atg7 knockout mice exhibited increased renal injury compared with wild-type mice following cisplatin- and ischemia- reperfusion injury-induced, suggesting a protective role of autophagy.

Hsp70i is required for autophagy due to many forms of stress (Yang Yonghua et al. 2013). In an immune context, the inducible Hsp70 chaperone is also associated with autophagy, leading to a preferential uploading of Hsp70i peptides, in major histocompatibility complex class II molecules of stressed cells. Then, Tregs, critical subtype of lymphocytes that display potent immunosuppressive effects, target the conserved Hsp70i peptides that may then be presented in these conditions leading to the repression of inflammation (Borges TJ et al. 2012).

Recently, it has been demonstrated that the protective effect of Hsp70i chaperone in AKI is Treg- dependent with the influence of the chaperone on its behavior, possibly being driven through splenic $CD11c^+$ dendritic cells (Kim et al. 2014). O'Neill et al. used Treg depletion strategies to show that heat preconditioning leads to inducible Hsp70 chaperone and mediated protection of kidney function from renal acute injury being Treg-dependent (O'Neill et al. 2014) (201).

During thermal or other stressors, wild type Hsp70 rapidly translocates to the nucleus to stabilize histones (Vanderwaal et al. 2009). Hsp70 also accumulates in the cytosol to ameliorate protein denaturation and promote multimeric protein complex assembly during stress (Rosenzweig et al. 2019) (203).

In ischemic AKI, nucleophosmin (NPM) and Bax participate in the regulation of cell death that begins only when both partners occupy the same cell compartment and are conformationally altered to become suitable binding partners. This protein:protein interaction allows the NPM-Bax complex to traffic to mitochondria, permeabilize the outer membrane injury, and release cell death mediators (Kerr et al. 2007; Wang Zhiyong et al. 2019).

After ischemic stress, Wang et al., demonstrated the Hsp70 significantly increases cell survival and decreases apoptosis by inhibiting NPM monomer translocation along with NPM less available to complex with cytosolic Bax, traffic to mitochondria and release cell death mediators (Wang Zhiyong et al. 2022). These findings will provide stimulus to further examine the role and mechanism of the enhancement of autophagy in acute kidney injury.

Involvement of Heat Shock Protein 70 in the Antioxidative Action of Losartan in Proximal Tubule Epithelial Cells

We investigated the contributions of caveolin 1 and the inducible Hsp70 to the regulation of the NADPH oxidase 4 (Nox4) expression in the proximal tubules of SHR. Due to the interaction of caveolin 1 and Hsp70i after angiotensin II AT_1 receptor blockade, a role of both proteins in AT_1 regulation was demonstrated in cell membranes from SHR microdissected proximal tubules. Plasma membrane translocated Hsp70i and its colocalization with caveolin 1 could be involved in the mechanism responsible for losartan cytoprotective effect on proximal tubules by decreasing oxidative stress through the downregulation of NADPH oxidase subunit Nox4. In coprecipitation experiments, increased amounts of Hsp70i were shown to interact with decreased amounts of Nox4 subunits, revealing a physical and functional association between both proteins (Bocanegra et al. 2010). Supporting our results, a functional interaction of HSPA1A with NAD(P)H: quinone oxidoreductase 1 (NQO1), a potent antioxidant that catalyzes two-electron reduction of various quinones, with NADH or NADPH as an electron donor, was previously found in coimmunoprecipitation; interaction of the chaperone could only be observed with wild-type NQO1*1 but not with the mutant NQO1*2, a point mutation in exon 6 of the human NQO1 gene. The association of inducible Hsp70 chaperone with NQO1*1 may play an important role in the stability and functionality of the NQO1 protein (Anwar et al. 2002). Our results allowed us to suggest that Hsp70i mediated decrease in the expression of Nox4 in isolated cell membranes of losartan-treated SHR microdissected proximal tubules, probably inducing the translocation of Nox4 into small intracellular vesicles (early or late endosomes), in route to their degradation (Bocanegra et al. 2010).In addition, in SHR vascular smooth muscle cells (VSMCs), we reported on how a functional association of the Hsp70i with Nox4/p22 phox provides a mechanism, which explains an antioxidative effect of losartan along with subsequent cytoskeletal integrity modulation (Gil Lorenzo et al. 2014).

Afterwards, we investigated how the inducible Hsp70 chaperone regulates Nox4 subunit trafficking and degradation in losartan-treated SHR PTECs. It is known that Hsp70 promotes the proteasomal degradation of proteins by recruiting the ubiquitin ligase CHIP, a key component of the chaperone-dependent quality control mechanism (Wickner, Maurizi, and Gottesman 1999). Client proteins are efficiently targeted by CHIP, particularly when they are partially or frankly misfolded, as is the case for most proteins binding to Hsp70 through exposed hydrophobic residues (Demand et al. 2001). In this way, Nox4 is roughly divided in two large domains: an N-terminal cluster of hydrophobic membrane-spanning sequences, and a C-terminal protein domain (Cheng et al. 2001).

We demonstrated that losartan induces increased Hsp70i expression and decreased oxidative stress through Nox4 protein levels reduction in PTECs from SHRs. The decreased Nox4 was not due to modifications at transcriptional level expression, it was as a result of the interaction of Hsp70 with its CHIP cochaperone complex that negatively regulates Nox4 protein levels through proteasomal degradation. Conversely, Hsp70i knockdown enhanced Nox4 protein levels, NADPH oxidase activity, and ROS generation in SHR PTECs, revealing that losartan was unable to abrogate angiotensin II effects on Nox4 expression and oxidative stress activity (Lorenzo et al. 2015).

Taken together, these losses of function data indicate that the inducible Hsp70 chaperone is directly involved in Nox4 regulation through the physical and functional interactions that come under the antioxidative effect of losartan, Figure 3 (Lorenzo et al. 2015).

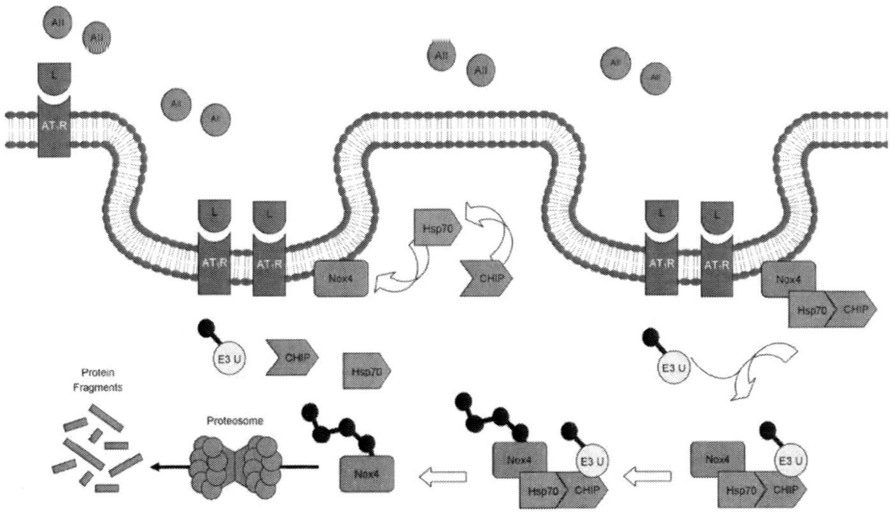

Figure 3. Mechanism by which the inducible Hsp70 chaperone/CHIP interaction mediates the negative regulation of Nox4 within the antioxidative effect of losartan in proximal tubule cells from spontaneously hypertensive rats (SHRs).

Angiotensin II and Epithelial-Mesenchymal Transition: Role of Hsp70

The main function of the epithelial cell is to provide structure and prevent permeability; therefore, its most prominent characteristic is tight junctions and polarity. On the other hand, the prototype for mesenchymal cells is fibroblasts whose function is to deposit collagen as a repair mechanism often secondary to an insult. EMT is a phenomenon where inherent epithelial markers are lost and mesenchymal characteristics are gained. Simplistically speaking, an epithelial cell transitions into a mesenchymal cell, losing all the characteristics inherent to epithelial cells and gaining mesenchymal features in the process. The end result is a motile, collagen-producing cell from epithelial origin (Kalluri and Weinberg 2009). In recent years EMT has been found to play a crucial role in fibrosis, in particular in the kidney (Hills and Squires 2010). The main inducers of EMT are TGF-β and angiotensin II, the same factors that classically induce fibrosis, suggesting that EMT is a common mechanism underlying TIF (Rodrigues-Díez et al. 2008).

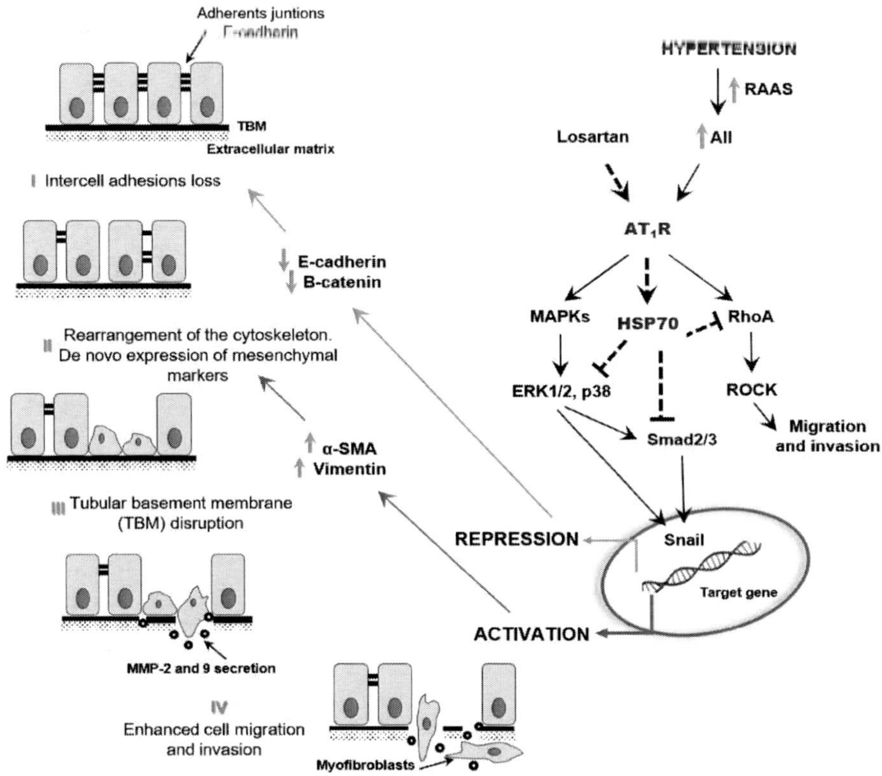

Figure 4. Proposed mechanism by which losartan through Hsp70 inhibits the epithelial to mesenchymal transition (EMT) induced by angiotensin II in proximal tubule cells derived from SHR.
Left panel: schematic illustration showing four key events during tubular epithelial cells to myofibroblast transition.
Right panel: throughout the activation of the MAPK cascade and the RhoA/ROCK pathway, angiotensin II regulates EMT markers. The Hps70 chaperone inhibits the signaling pathways that lead to epithelial-mesenchymal transition after losartan AT_1 receptor blockage.

We focused on the role of Hsp70 through losartan effect on angiotensin II induced epithelial–mesenchymal transition in primary culture of PTECs from a genetic experimental model of hypertension (SHR). We demonstrate that Hsp70i within the losartan effect inhibits the EMT induced by angiotensin II in PTECs from SHR. Losartan increased E-cadherin levels and decreased vinculin, α-SMA, vimentin, pERK, p38 and Smad2–3 activation compared to angiotensin II treated and without treatment SHR PTECs. Moreover, losartan inhibited matrix metalloproteinase-2 and 9 secretion, reduced migration and cellular displacement, stabilizing intercellular junctions.

Notably, losartan treatment in shHsp72 knockdown SHR PTECs showed results similar to SHR PTECs that lead to the EMT process. Losartan treatment in SHR PTECs induced decrease in Rac1 and RhoA activation. Although, in shHsp72 Knockdown SHR PTECs no significant changes were observed (Costantino et al. 2019). According to these results, Rodrigues-Díez et al. 217 demonstrated that RhoA inhibition through transient transfection of a dominant negative RhoA or the use of ROCK inhibitors regulates angiotensin II-mediated EMT in angiotensin II treated human proximal tubule cells (HK2).

In our study, we demonstrated that losartan avoids the epithelial mesenchymal transition induced by angiotensin II, and we characterize the Hsp70 chaperone as an important mediator of the losartan effect which entails migration reduction and cytoskeletal stabilization in proximal tubule cells (Figure 4). We focus on this chaperone as a potential target that ameliorates renal damage.

Our data indicate that Hsp70 is involved in the cellular response after AT_1 receptor blockage, inducing decreased profibrotic factors that lead to mesenchymal–epithelial transition.

Conclusion

Tubulointerstitial fibrosis represents the final common pathway of any form of progressive renal disease. The severity of tubular interstitial damage is highly correlated to the degree of decline of renal function, even better than the glomerular lesions do. Increased attention has focused on how the responses of proximal tubule to injury may mediate TIF progression.

Many of these proximal tubular responses are initially adaptive, but as the injury persists, these responses may promote tubular atrophy and TIF through autocrine and paracrine signaling pathways. Renal local renin- angiotensin system (RAS) activation plays a pivotal role in CKD progression. Angiotensin II has been directly implicated in tubulointerstitial injury and fibrosis. Blockade of the RAS has become the mainstay therapy for the preservation of CKD. So far, epithelial–mesenchymal transition has emerged as a main contributor to fibrosis in end-stage organ disease and, as such, may become a new potential anti-fibrotic target.

Our data suggests a key role of Hsp70 chaperone on the angiotensin II–induced EMT after AT_1 receptor blockage, this being a possible molecular target for therapeutic strategy against TIF.

Hence, a better understanding of the mechanisms by which proximal tubule injury drives inflammation and fibrosis will undoubtedly help the development of novel therapies to halt renal tubulointerstitial fibrosis and CKD progression.

References

Assimon, Victoria, Anne T. Gillies, Jennifer N. Rauch, and Jason E. Gestwicki. 2013. "Hsp70 Protein Complexes as Drug Targets." *Current Pharmaceutical Design* 19 (3): 404–17. https://doi.org/10.2174/138161213804143699.

Abrams, Jennifer L., Jacob Verghese, Patrick A. Gibney, and Kevin A. Morano. 2014. "Hierarchical Functional Specificity of Cytosolic Heat Shock Protein 70 (Hsp70) Nucleotide Exchange Factors in Yeast." *The Journal of Biological Chemistry* 289 (19): 13155–67. https://doi.org/10.1074/JBC.M113.530014.

Alderson, T. Reid, Jin Hae Kim, Kai Cai, Ronnie O. Frederick, Marco Tonelli, and John L. Markley. 2014. "The Specialized Hsp70 (HscA) Interdomain Linker Binds to Its Nucleotide-Binding Domain and Stimulates ATP Hydrolysis in Both Cis and Trans Configurations." *Biochemistry* 53 (46): 7148–59. https://doi.org/10.1021/BI5010552.

Allam, Ramanjaneyulu, Christina Rebecca Scherbaum, Murthy Narayana Darisipudi, Shrikant R Mulay, Holger Hägele, Julia Lichtnekert, Jan Henrik Hagemann, Khader Valli Rupanagudi, Mi Ryu, Claudia Schwarzenberger, Bernd Hohenstein, Christian Hugo, Bernd Uhl, Christoph A Reichel, Fritz Krombach, Marc Monestier, Helen Liapis, Kristin Moreth, Liliana Schaefer, Hans-Joachim Anders. 2012. "Histones from Dying Renal Cells Aggravate Kidney Injury via TLR2 and TLR4." *Journal of the American Society of Nephrology* 23 (8): 1375–88. https://doi.org/10.1681/ASN.2011111077/-/DCSUPPLEMENTAL.

Al-Lamki, Rafia S., and Tanya N. Mayadas. 2015. "TNF Receptors: Signaling Pathways and Contribution to Renal Dysfunction." *Kidney International* 87 (2): 281–96. https://doi.org/10.1038/KI.2014.285.

Anders, Hans Joachim. 2016. "Of Inflammasomes and Alarmins: IL-1β and IL-1α in Kidney Disease." *Journal of the American Society of Nephrology JASN* 27 (9): 2564–75. https://doi.org/10.1681/ASN.2016020177.

Anders, Hans Joachim, and Liliana Schaefer. 2014. "Beyond Tissue Injury-Damage-Associated Molecular Patterns, Toll-like Receptors, and Inflammasomes Also Drive Regeneration and Fibrosis." *Journal of the American Society of Nephrology: JASN* 25 (7): 1387–1400. https://doi.org/10.1681/ASN.2014010117.

Anwar, Adil, David Siegel, Jadwiga K. Kepa, and David Ross. 2002. "Interaction of the Molecular Chaperone Hsp70 with Human NAD(P)H:Quinone Oxidoreductase 1." *The*

Journal of Biological Chemistry 277 (16): 14060–67. https://doi.org/10.1074/JBC.M111576200.

Aufricht, Christoph, Ellen Lu, Gunilla Thulin, Michael Kashgarian, Norman J. Siegel, and Scott K. Van Why. 1998. "ATP Releases HSP-72 from Protein Aggregates after Renal Ischemia." *The American Journal of Physiology* 274 (2). https://doi.org/10.1152/AJPRENAL.1998.274.2.F268.

Balchin, David, Manajit Hayer-Hartl, and F. Ulrich Hartl. 2016. "In Vivo Aspects of Protein Folding and Quality Control." *Science (New York, N.Y.)* 353 (6294). https://doi.org/10.1126/SCIENCE.AAC4354.

Ballinger, Carol A., Patrice Connell, Yaxu Wu, Zhaoyong Hu, Larry J. Thompson, Li-Yan Yin, and Cam Patterson. 1999. "Identification of CHIP, a Novel Tetratricopeptide Repeat-Containing Protein That Interacts with Heat Shock Proteins and Negatively Regulates Chaperone Functions." *Molecular and Cellular Biology* 19 (6): 4535–45. https://doi.org/10.1128/MCB.19.6.4535.

Beck, Franz X., Wolfgang Neuhofer, and Eva Müller. 2000. "Molecular Chaperones in the Kidney: Distribution, Putative Roles, and Regulation." *American Journal of Physiology. Renal Physiology* 279 (2). https://doi.org/10.1152/AJPRENAL.2000.279.2.F203.

Bocanegra, Victoria, Walter Manucha, Marcelo Rodríguez Peña, Valeria Cacciamani, and Patricia G. Vallés. 2010. "Caveolin-1 and Hsp70 Interaction in Microdissected Proximal Tubules from Spontaneously Hypertensive Rats as an Effect of Losartan." *Journal of Hypertension* 28 (1): 143–55. https://doi.org/10.1097/HJH.0B013E328332B778.

Bohle, A., H. Christ, K. E. Grund, and S. Mackensen. 1979. "The Role of the Interstitium of the Renal Cortex in Renal Disease." *Contributions to Nephrology* 16: 109–14. https://doi.org/10.1159/000402883.

Bohrer, M. P., W. M. Deen, C. R. Robertson, and B. M. Brenner. 1977. "Mechanism of Angiotensin II-Induced Proteinuria in the Rat." *The American Journal of Physiology* 233 (1). https://doi.org/10.1152/AJPRENAL.1977.233.1.F13.

Boor, Peter, Tammo Ostendorf, and Jürgen Floege. 2010. "Renal Fibrosis: Novel Insights into Mechanisms and Therapeutic Targets." *Nature Reviews. Nephrology* 6 (11): 643–56. https://doi.org/10.1038/NRNEPH.2010.120.

Borges, Fernanda T., Sonia A. Melo, Berna C. Özdemir, Noritoshi Kato, Ignacio Revuelta, Caroline A. Miller, Vincent H. Gattone, Valerie S. LeBleu, and Raghu Kalluri. 2013. "TGF-B1-Containing Exosomes from Injured Epithelial Cells Activate Fibroblasts to Initiate Tissue Regenerative Responses and Fibrosis." *Journal of the American Society of Nephrology: JASN* 24 (3): 385–92. https://doi.org/10.1681/ASN.2012101031.

Borges, Thiago J., Lotte Wieten, Martijn J. C. Van Herwijnen, Femke Broere, Ruurd Van derZee, Cristina Bonorino, and Willem Van Eden. 2012. "The Anti-Inflammatory Mechanisms of Hsp70." *Frontiers in Immunology* 3 (MAY). https://doi.org/10.3389/FIMMU.2012.00095.

Bracher, Andreas, and Jacob Verghese. 2015. "The Nucleotide Exchange Factors of Hsp70 Molecular Chaperones." *Frontiers in Molecular Biosciences* 2 (APR). https://doi.org/10.3389/FMOLB.2015.00010.

Casarini, D. E., M. A. Boim, R. C. R. Stella, M. H. Krieger-Azzolini, J. E. Krieger, and N. Schor. 1997. "Angiotensin I-Converting Enzyme Activity in Tubular Fluid along the Rat Nephron." *The American Journal of Physiology* 272 (3 Pt 2). https://doi.org/10.1152/AJPRENAL.1997.272.3.F405.

Chappell, Mark C. 2012. "Nonclassical Renin-Angiotensin System and Renal Function." *Comprehensive Physiology* 2 (4): 2733–52. https://doi.org/10.1002/CPHY.C120002.

Chebotareva, Natalia, Irina Bobkova, and Evgeniy Shilov. 2017. "Heat Shock Proteins and Kidney Disease: Perspectives of HSP Therapy." *Cell Stress & Chaperones* 22 (3): 319–43. https://doi.org/10.1007/S12192-017-0790-0.

Chen, Y., C. Yang, Z. Xie, L. Zou, Z. Ruan, X. Zhang, Y. Tang, L. Fei, Z. Jia, and Y. Wu. 2006. "Expression of the Novel Co-Stimulatory Molecule B7-H4 by Renal Tubular Epithelial Cells." *Kidney International* 70 (12): 2092–99. https://doi.org/10.1038/SJ.KI.5001867.

Chen, Yi, Fan-Chi Chang, Ching-Fang Wu, Yu-Hsiang Chou, Huan-Lun Hsu, Wen-Chih Chiang, Juqun Shen, Yung-Ming Chen, Kwan-Dun Wu, Tun-Jun Tsai, Jeremy S Duffield, Shuei-Liong Lin. 2011. "Platelet-Derived Growth Factor Receptor Signaling Activates Pericyte-Myofibroblast Transition in Obstructive and Post-Ischemic Kidney Fibrosis." *Kidney International* 80 (11): 1170–81. https://doi.org/10.1038/KI.2011.208.

Cheng, Guangjie, Zehong Cao, Xiangxi Xu, Erwin G.Van Meir, and J. David Lambeth. 2001. "Homologs of Gp91phox: Cloning and Tissue Expression of Nox3, Nox4, and Nox5." *Gene* 269 (1–2): 131–40. https://doi.org/10.1016/S0378-1119(01)00449-8.

Chung, Arthur C. K., and Hui Y. Lan. 2011. "Chemokines in Renal Injury." *Journal of the American Society of Nephrology: JASN* 22 (5): 802–9. https://doi.org/10.1681/ASN.2010050510.

Costantino, Valeria V., Victoria Bocanegra, Valeria Cacciamani, Andrea F. Gil Lorenzo, María E. Benardon, and Patricia G. Vallés. 2019. "Losartan through Hsp70 Avoids Angiotensin II Induced Mesenchymal Epithelial Transition in Proximal Tubule Cells from Spontaneously Hypertensive Rats." *Cellular Physiology and Biochemistry: International Journal of Experimental Cellular Physiology, Biochemistry, and Pharmacology* 53 (4): 713–30. https://doi.org/10.33594/000000167.

Craig, Elizabeth A. 2018. "Hsp70 at the Membrane: Driving Protein Translocation." *BMC Biology* 16 (1). https://doi.org/10.1186/S12915-017-0474-3.

Dear, J. W., H. Yasuda, X. Hu, S. Hieny, P. S. T. Yuen, S. M. Hewitt, A. Sher, and R. A. Star. 2006. "Sepsis-Induced Organ Failure Is Mediated by Different Pathways in the Kidney and Liver: Acute Renal Failure Is Dependent on MyD88 but Not Renal Cell Apoptosis." *Kidney International* 69 (5): 832–36. https://doi.org/10.1038/SJ.KI.5000165.

Demand, Jens, Simon Alberti, Cam Patterson, and Jörg Höhfeld. 2001. "Cooperation of a Ubiquitin Domain Protein and an E3 Ubiquitin Ligase during Chaperone/Proteasome Coupling." *Current Biology: CB* 11 (20): 1569–77. https://doi.org/10.1016/S0960-9822(01)00487-0.

Ding, Hong, Ying Zhou, and Haihua Huang. 2019. "MiR-101a Ameliorates AngII-Mediated Hypertensive Nephropathy by Blockade of TGFβ/Smad3 and NF-KB Signalling in a Mouse Model of Hypertension." *Clinical and Experimental*

Pharmacology & Physiology 46 (3): 246–54. https://doi.org/10.1111/1440-1681.13042.

Ding, Li Hong, Dan Liu, Min Xu, Min Wu, Hong Liu, Ri Ning Tang, Kun Ling Ma, Ping Sheng Chen, and Bi Cheng Liu. 2015. "TLR2-MyD88-NF-KB Pathway Is Involved in Tubulointerstitial Inflammation Caused by Proteinuria." *The International Journal of Biochemistry & Cell Biology* 69 (December): 114–20. https://doi.org/10.1016/J.BIOCEL.2015.10.014.

Durvasula, Raghu V., Arndt T. Petermann, Keiju Hiromura, Mary Blonski, Jeffrey Pippin, Peter Mundel, Raimund Pichler, Sian Griffin, William G. Couser, and Stuart J. Shankland. 2004. "Activation of a Local Tissue Angiotensin System in Podocytes by Mechanical Strain." *Kidney International* 65 (1): 30–39. https://doi.org/10.1111/J.1523-1755.2004.00362.X.

Ebisawa, Takanori, Minoru Fukuchi, Gyo Murakami, Tomoki Chiba, Keiji Tanaka, Takeshi Imamura, and Kohei Miyazono. 2001. "Smurf1 Interacts with Transforming Growth Factor-Beta Type I Receptor through Smad7 and Induces Receptor Degradation." *The Journal of Biological Chemistry* 276 (16): 12477–80. https://doi.org/10.1074/JBC.C100008200.

El-Achkar, Tarek M., and Pierre C. Dagher. 2006. "Renal Toll-like Receptors: Recent Advances and Implications for Disease." *Nature Clinical Practice. Nephrology* 2 (10): 568–81. https://doi.org/10.1038/NCPNEPH0300.

El-Achkar, Tarek M., Xiaoping Huang, Zoya Plotkin, Ruben M. Sandoval, Georges J. Rhodes, and Pierre C. Dagher. 2006. "Sepsis Induces Changes in the Expression and Distribution of Toll-like Receptor 4 in the Rat Kidney." *American Journal of Physiology. Renal Physiology* 290 (5). https://doi.org/10.1152/AJPRENAL.00414.2005.

Emami, A., J. H. Schwartz, and S. C. Borkan. 1991. "Transient Ischemia or Heat Stress Induces a Cytoprotectant Protein in Rat Kidney." *The American Journal of Physiology* 260 (4 Pt 2). https://doi.org/10.1152/AJPRENAL.1991.260.4.F479.

Ferenbach, David A., and Joseph V. Bonventre. 2015. "Mechanisms of Maladaptive Repair after AKI Leading to Accelerated Kidney Ageing and CKD." *Nature Reviews. Nephrology* 11 (5): 264–76. https://doi.org/10.1038/NRNEPH.2015.3.

Fernández-Fernández, María Rosario, Marcos Gragera, Lissette Ochoa-Ibarrola, Lucía Quintana-Gallardo, and José María Valpuesta. 2017. "Hsp70 - a Master Regulator in Protein Degradation." *FEBS Letters* 591 (17): 2648–60. https://doi.org/10.1002/1873-3468.12751.

Fine, Leon G., and Jill T. Norman. 2008. "Chronic Hypoxia as a Mechanism of Progression of Chronic Kidney Diseases: From Hypothesis to Novel Therapeutics." *Kidney International* 74 (7): 867–72. https://doi.org/10.1038/KI.2008.350.

Fintha, A., A. Sebe, A. Masszi, T. Terebessy, T. Huszár, L. Rosivall, and Istvan Mucsi. 2007. "Angiotensin II Activates Plasminogen Activator Inhibitor-I Promoter in Renal Tubular Epithelial Cells via the AT1 Receptor." *Acta Physiologica Hungarica* 94 (1–2): 19–30. https://doi.org/10.1556/APHYSIOL.94.2007.1-2.4.

Fragiadaki, Maria, Abigail S. Witherden, Tomoyo Kaneko, Sonali Sonnylal, Charles D. Pusey, George Bou-Gharios, and Roger M. Mason. 2011. "Interstitial Fibrosis Is Associated with Increased COL1A2 Transcription in AA-Injured Renal Tubular

Epithelial Cells in Vivo." *Matrix Biology: Journal of the International Society for Matrix Biology* 30 (7–8): 396–403. https://doi.org/10.1016/J.MATBIO.2011.07.004.

Freedman, Barry I., and Arthur H. Cohen. 2016. "Hypertension-Attributed Nephropathy: What's in a Name?" *Nature Reviews. Nephrology* 12 (1): 27–36. https://doi.org/10.1038/NRNEPH.2015.172.

Furuichi, Kengo, Takashi Wada, Yasunori Iwata, Kiyoki Kitagawa, Ken Ichi Kobayashi, Hiroyuki Hashimoto, Yoshiro Ishiwata, Masahide Asano, Hui Wang, Kouji Matsushima, Motohiro Takeya, William A Kuziel, Naofumi Mukaida, Hitoshi Yokoyama. 2003. "CCR2 Signaling Contributes to Ischemia-Reperfusion Injury in Kidney." *Journal of the American Society of Nephrology: JASN* 14 (10): 2503–15. https://doi.org/10.1097/01.ASN.0000089563.63641.A8.

Gabriel Navar, L., and Akira Nishiyama. 2001. "Intrarenal Formation of Angiotensin II." *Contributions to Nephrology* 135 (135): 1–15. https://doi.org/10.1159/000060154.

Gewin, Leslie S. 2018. "Renal Fibrosis: Primacy of the Proximal Tubule." *Matrix Biology: Journal of the International Society for Matrix Biology* 68–69 (August): 248–62. https://doi.org/10.1016/J.MATBIO.2018.02.006.

Gil Lorenzo, Andrea Fernanda, Victoria Bocanegra, María Eugenia Benardon, Valeria Cacciamani, and Patricia G. Vallés. 2014. "Hsp70 Regulation on Nox4/P22phox and Cytoskeletal Integrity as an Effect of Losartan in Vascular Smooth Muscle Cells." *Cell Stress & Chaperones* 19 (1): 115–34. https://doi.org/10.1007/S12192-013-0439-6.

Gill, Pritmohinder S., and Christopher S. Wilcox. 2006. "NADPH Oxidases in the Kidney." *Antioxidants & Redox Signaling* 8 (9–10): 1597–1607. https://doi.org/10.1089/ARS.2006.8.1597.

Gluba, Anna, MacIej Banach, Simon Hannam, Dimitri P. Mikhailidis, Agata Sakowicz, and Jacek Rysz. 2010. "The Role of Toll-like Receptors in Renal Diseases." *Nature Reviews. Nephrology* 6 (4): 224–35. https://doi.org/10.1038/NRNEPH.2010.16.

Gomez-Pastor, Rocio, Eileen T. Burchfiel, and Dennis J. Thiele. 2018. "Regulation of Heat Shock Transcription Factors and Their Roles in Physiology and Disease." *Nature Reviews. Molecular Cell Biology* 19 (1): 4–19. https://doi.org/10.1038/NRM.2017.73.

Gonçalves, Giselle Martins, Dario S. Zamboni, and Niels Olsen Saraiva Cĝmara. 2010. "The Role of Innate Immunity in Septic Acute Kidney Injuries." *Shock (Augusta, Ga.)* 34 Suppl 1 (SUPPL. 1): 22–26. https://doi.org/10.1097/SHK.0B013E3181E7E69E.

Grande, M. Teresa, Berta Sánchez-Laorden, Cristina López-Blau, Cristina A. De Frutos, Agnès Boutet, Miguel Arévalo, R. Grant Rowe, Stephen J. Weiss, José M. López-Novoa, and M. Angela Nieto. 2015. "Snail1-Induced Partial Epithelial-to-Mesenchymal Transition Drives Renal Fibrosis in Mice and Can Be Targeted to Reverse Established Disease." *Nature Medicine* 21 (9): 989–97. https://doi.org/10.1038/NM.3901.

Harris, P. J., and L. G. Navar. 1985. "Tubular Transport Responses to Angiotensin." *The American Journal of Physiology* 248 (5 Pt 2). https://doi.org/10.1152/AJPRENAL.1985.248.5.F621.

Haugen, Eric N., Anthony J. Croatt, and Karl A. Nath. 2000. "Angiotensin II Induces Renal Oxidant Stress in Vivo and Heme Oxygenase-1 in Vivo and in Vitro." *Kidney International* 58 (1): 144–52. https://doi.org/10.1046/J.1523-1755.2000.00150.X.

Heyman, Samuel N., Mogher Khamaisi, Seymour Rosen, and Christian Rosenberger. 2008. "Renal Parenchymal Hypoxia, Hypoxia Response and the Progression of Chronic Kidney Disease." *American Journal of Nephrology* 28 (6): 998–1006. https://doi.org/10.1159/000146075.

Hill, Gary S. 2008. "Hypertensive Nephrosclerosis." *Current Opinion in Nephrology and Hypertension* 17 (3): 266–70. https://doi.org/10.1097/MNH.0B013E3282F88A1F.

Hill, Nathan R., Samuel T. Fatoba, Jason L. Oke, Jennifer A. Hirst, Christopher A. O'Callaghan, Daniel S. Lasserson, and F. D.Richard Hobbs. 2016. "Global Prevalence of Chronic Kidney Disease - A Systematic Review and Meta-Analysis." *PloS One* 11 (7). https://doi.org/10.1371/JOURNAL.PONE.0158765.

Hills, Claire E., and Paul E. Squires. 2010. "TGF-Beta1-Induced Epithelial-to-Mesenchymal Transition and Therapeutic Intervention in Diabetic Nephropathy." *American Journal of Nephrology* 31 (1): 68–74. https://doi.org/10.1159/000256659.

Hitomi, Hirofumi, Hideyasu Kiyomoto, and Akira Nishiyama. 2007. "Angiotensin II and Oxidative Stress." *Current Opinion in Cardiology* 22 (4): 311–15. https://doi.org/10.1097/HCO.0B013E3281532B53.

Huang, Xiao R., Arthur C. K. Chung, Fuye Yang, Wensheng Yue, Chuxia Deng, Chu Pak Lau, Hung Fat Tse, and Hui Y. Lan. 2010. "Smad3 Mediates Cardiac Inflammation and Fibrosis in Angiotensin II-Induced Hypertensive Cardiac Remodeling." *Hypertension* (Dallas, Tex.: 1979) 55 (5): 1165–71. https://doi.org/10.1161/HYPERTENSIONAHA.109.147611.

Ishizaka, Nobukazu, Kathy K. Griendling, Bernard Lassègue, and R. Wayne Alexander. 1998. "Angiotensin II Type 1 Receptor: Relationship with Caveolae and Caveolin after Initial Agonist Stimulation." *Hypertension (Dallas, Tex.: 1979)* 32 (3): 459–66. https://doi.org/10.1161/01.HYP.32.3.459.

Iwano, Masayuki, David Plieth, Theodore M. Danoff, Chengsen Xue, Hirokazu Okada, and Eric G. Neilson. 2002. "Evidence That Fibroblasts Derive from Epithelium during Tissue Fibrosis." *The Journal of Clinical Investigation* 110 (3): 341–50. https://doi.org/10.1172/JCI15518.

Jevnikar, Anthony M., Daniel C. Brennan, Gary G. Singer, John E. Heng, Wowo Maslinski, Rudolf P. Wuthrich, Laurie H. Glimcher, and Vicki E. Rubin Kelley. 1991. "Stimulated Kidney Tubular Epithelial Cells Express Membrane Associated and Secreted TNF Alpha." *Kidney International* 40 (2): 203–11. https://doi.org/10.1038/KI.1991.201.

Kalluri, Raghu, and Robert A. Weinberg. 2009. "The Basics of Epithelial-Mesenchymal Transition." *The Journal of Clinical Investigation* 119 (6): 1420–28. https://doi.org/10.1172/JCI39104.

Kampinga, Harm H., Jurre Hageman, Michel J. Vos, Hiroshi Kubota, Robert M. Tanguay, Elspeth A. Bruford, Michael E. Cheetham, Bin Chen, and Lawrence E. Hightower. 2009. "Guidelines for the Nomenclature of the Human Heat Shock Proteins." *Cell Stress & Chaperones* 14 (1): 105–11. https://doi.org/10.1007/S12192-008-0068-7.

Kang, Duk-Hee, Alison H. Joly, Se-Woong Oh, Christian Hugo, Dontscho Kerjaschki, Katherine l. Gordon, Marilda Mazzali, Ashley Jefferson, Jeremy Hughes, Kirsten M Madsen, George F Schreiner, Richard J Johnson. 2001. "Impaired Angiogenesis in the Remnant Kidney Model: I. Potential Role of Vascular Endothelial Growth Factor and

Thrombospondin-1." *Journal of the American Society of Nephrology: JASN* 12 (7): 1434–47. https://doi.org/10.1681/ASN.V1271434.

Kang, Hyun Mi, Seon Ho Ahn, Peter Choi, Yi An Ko, Seung Hyeok Han, Frank Chinga, Ae Seo Deok Park, Jianling Tao, Kumar Sharma, James Pullman, Erwin P Bottinger, Ira J Goldberg, Katalin Susztak. 2015. "Defective Fatty Acid Oxidation in Renal Tubular Epithelial Cells Has a Key Role in Kidney Fibrosis Development." *Nature Medicine* 21 (1): 37–46. https://doi.org/10.1038/NM.3762.

Kaushal, Gur P. 2012. "Autophagy Protects Proximal Tubular Cells from Injury and Apoptosis." *Kidney International* 82 (12): 1250–53. https://doi.org/10.1038/KI.2012.337.

Kavsak, Peter, Richele K. Rasmussen, Carrie G. Causing, Shirin Bonni, Haitao Zhu, Gerald H. Thomsen, and Jeffrey L. Wrana. 2000. "Smad7 Binds to Smurf2 to Form an E3 Ubiquitin Ligase That Targets the TGF Beta Receptor for Degradation." *Molecular Cell* 6 (6): 1365–75. https://doi.org/10.1016/S1097-2765(00)00134-9.

Kelley, V R, and G C Singer. 1993. "The Antigen Presentation Function of Renal Tubular Epithelial Cells." *Exp Nephrol.* March 1993. https://pubmed.ncbi.nlm.nih.gov/8081954/.

Kelley, William L. 1998. "The J-Domain Family and the Recruitment of Chaperone Power." *Trends in Biochemical Sciences* 23 (6): 222–27. https://doi.org/10.1016/S0968-0004(98)01215-8.

Kerr, L. E., J. L. A. Birse-Archbold, D. M. Short, A. L. McGregor, I. Heron, D. C. MacDonald, J. Thompson, G J Carlson, J S Kelly, J McCulloch, J Sharkey. 2007. "Nucleophosmin Is a Novel Bax Chaperone That Regulates Apoptotic Cell Death." *Oncogene* 26 (18): 2554–62. https://doi.org/10.1038/SJ.ONC.1210044.

Kim, Myung Gyu, Eun Jung Cho, Jae Won Lee, Yoon Sook Ko, Hee Young Lee, Sang Kyung Jo, Won Yong Cho, and Hyoung Kyu Kim. 2014. "The Heat-Shock Protein-70-Induced Renoprotective Effect Is Partially Mediated by CD4+ CD25+ Foxp3 + Regulatory T Cells in Ischemia/Reperfusion-Induced Acute Kidney Injury." *Kidney International* 85 (1): 62–71. https://doi.org/10.1038/KI.2013.277.

Kobori, Hiroyuki, Yuri Ozawa, Ryousuke Satou, Akemi Katsurada, Kayoko Miyata, Naro Ohashi, Naoki Hase, Yuki Suzaki, Curt D. Sigmund, and L. Gabriel Navar. 2007. "Kidney-Specific Enhancement of ANG II Stimulates Endogenous Intrarenal Angiotensinogen in Gene-Targeted Mice." *American Journal of Physiology. Renal Physiology* 293 (3). https://doi.org/10.1152/AJPRENAL.00146.2007.

Koesters, Robert, Brigitte Kaissling, Michel LeHir, Nicolas Picard, Franziska Theilig, Rolf Gebhardt, Adam B. Glick, Brunhilde Hähnel, Hiltraud Hosser, Hermann-Josef Gröne, Wilhelm Kriz. 2010. "Tubular Overexpression of Transforming Growth Factor-Beta1 Induces Autophagy and Fibrosis but Not Mesenchymal Transition of Renal Epithelial Cells." *The American Journal of Pathology* 177 (2): 632–43. https://doi.org/10.2353/AJPATH.2010.091012.

Komatsuda, A., H. Wakui, H. Imai, Y. Nakamoto, A. B. Miura, H. Itoh, and Y. Tashima. 1992. "Renal Localization of the Constitutive 73-KDa Heat-Shock Protein in Normal and PAN Rats." *Kidney International* 41 (5): 1204–12. https://doi.org/10.1038/KI.1992.182.

Kurts, Christian, Ulf Panzer, Hans Joachim Anders, and Andrew J. Rees. 2013. "The Immune System and Kidney Disease: Basic Concepts and Clinical Implications." *Nature Reviews. Immunology* 13 (10): 738–53. https://doi.org/10.1038/NRI3523.

Lebleu, Valerie S., Gangadhar Taduri, Joyce O'Connell, Yingqi Teng, Vesselina G. Cooke, Craig Woda, Hikaru Sugimoto, and Raghu Kalluri. 2013. "Origin and Function of Myofibroblasts in Kidney Fibrosis." *Nature Medicine* 19 (8): 1047–53. https://doi.org/10.1038/NM.3218.

Leemans, Jaklien C., Lotte Kors, Hans Joachim Anders, and Sandrine Florquin. 2014. "Pattern Recognition Receptors and the Inflammasome in Kidney Disease." *Nature Reviews. Nephrology* 10 (7): 398–414. https://doi.org/10.1038/NRNEPH.2014.91.

Li, Hongye, and Edward P. Nord. 2002. "CD40 Ligation Stimulates MCP-1 and IL-8 Production, TRAF6 Recruitment, and MAPK Activation in Proximal Tubule Cells." *American Journal of Physiology. Renal Physiology* 282 (6). https://doi.org/10.1152/AJPRENAL.00291.2001.

Hongye Li, Edward P Nord.2005. "CD40/CD154 Ligation Induces Mononuclear Cell Adhesion to Human Renal Proximal Tubule Cells via Increased ICAM-1 Expression." *American Journal of Physiology. Renal Physiology* 289 (1). https://doi.org/10.1152/AJPRENAL.00317.2004.

Li, Shenyang, Nithya Mariappan, Judit Megyesi, Brian Shank, Krishnaswamy Kannan, Sue Theus, Peter M. Price, Jeremy S. Duffield, and Didier Portilla. 2013. "Proximal Tubule PPARα Attenuates Renal Fibrosis and Inflammation Caused by Unilateral Ureteral Obstruction." *American Journal of Physiology. Renal Physiology* 305 (5). https://doi.org/10.1152/AJPRENAL.00309.2013.

Li, Xiao C., Oscar A. Carretero, L. Gabriel Navar, and Jia L. Zhuo. 2006. "AT1 Receptor-Mediated Accumulation of Extracellular Angiotensin II in Proximal Tubule Cells: Role of Cytoskeleton Microtubules and Tyrosine Phosphatases." *American Journal of Physiology. Renal Physiology* 291 (2). https://doi.org/10.1152/AJPRENAL.00405.2005.

Li, Xiao C., L. Gabriel Navar, Yuan Shao, and Jia L. Zhuo. 2007. "Genetic Deletion of AT1a Receptors Attenuates Intracellular Accumulation of ANG II in the Kidney of AT1a Receptor-Deficient Mice." *American Journal of Physiology. Renal Physiology* 293 (2). https://doi.org/10.1152/AJPRENAL.00489.2006.

Li, Xiao C., and Jia L. Zhuo. 2014. "Mechanisms of AT1a Receptor-Mediated Uptake of Angiotensin II by Proximal Tubule Cells: A Novel Role of the Multiligand Endocytic Receptor Megalin." *American Journal of Physiology. Renal Physiology* 307 (2). https://doi.org/10.1152/AJPRENAL.00693.2013.

Liu, Bi Cheng, Tao Tao Tang, and Lin Li Lv. 2019. "How Tubular Epithelial Cell Injury Contributes to Renal Fibrosis." *Advances in Experimental Medicine and Biology* 1165: 233–52. https://doi.org/10.1007/978-981-13-8871-2_11.

Liu, Bi Cheng, Tao Tao Tang, Lin Li Lv, and Hui Yao Lan. 2018. "Renal Tubule Injury: A Driving Force toward Chronic Kidney Disease." *Kidney International* 93 (3): 568–79. https://doi.org/10.1016/J.KINT.2017.09.033.

Liu, Dan, Yi Wen, Tao Tao Tang, Lin Li Lv, Ri Ning Tang, Hong Liu, Kun Ling Ma, Steve D. Crowley, and Bi Cheng Liu. 2015. "Megalin/Cubulin-Lysosome-Mediated Albumin Reabsorption Is Involved in the Tubular Cell Activation of NLRP3

Inflammasome and Tubulointerstitial Inflammation." *The Journal of Biological Chemistry* 290 (29): 18018–28. https://doi.org/10.1074/JBC.M115.662064.

Liu, Dan, Min Xu, Li Hong Ding, Lin Li Lv, Hong Liu, Kun Ling Ma, Ai Hua Zhang, Steven D. Crowley, and Bi Cheng Liu. 2014. "Activation of the Nlrp3 Inflammasome by Mitochondrial Reactive Oxygen Species: A Novel Mechanism of Albumin-Induced Tubulointerstitial Inflammation." *The International Journal of Biochemistry & Cell Biology* 57: 7–19. https://doi.org/10.1016/J.BIOCEL.2014.09.018.

Liu, Fang, Chih Chang Wei, Shyh Jong Wu, Isabelle Chenier, Shao Ling Zhang, Janos G. Filep, Julie R. Ingelfinger, and John S. D. Chan. 2009. "Apocynin Attenuates Tubular Apoptosis and Tubulointerstitial Fibrosis in Transgenic Mice Independent of Hypertension." *Kidney International* 75 (2): 156–66. https://doi.org/10.1038/KI.2008.509.

Liu, Fu You, Xiao Zhao Li, You Ming Peng, Hong Liu, and Ying Hong Liu. 2007. "Arkadia-Smad7-Mediated Positive Regulation of TGF-Beta Signaling in a Rat Model of Tubulointerstitial Fibrosis." *American Journal of Nephrology* 27 (2): 176–83. https://doi.org/10.1159/000100518.

Liu, Guan Xian, You Qi Li, Xiao R. Huang, Li Hua Wei, Yang Zhang, Min Feng, Xiao Ming Meng, Hai Yong Chen, Yong Jun Shi, and Hui Y. Lan. 2014. "Smad7 Inhibits AngII-Mediated Hypertensive Nephropathy in a Mouse Model of Hypertension." *Clinical Science* (London, England: 1979) 127 (3): 195–208. https://doi.org/10.1042/CS20130706.

Liu, Guan Xian, You Qi Li, Xiao R. Huang, Lihua Wei, Hai Yong Chen, Yong Jun Shi, Rainer L. Heuchel, and Hui Y. Lan. 2013. "Disruption of Smad7 Promotes ANG II-Mediated Renal Inflammation and Fibrosis via Sp1-TGF-β/Smad3-NF.KB-Dependent Mechanisms in Mice." *PloS One* 8 (1). https://doi.org/10.1371/JOURNAL.PONE.0053573.

Liu, Zhen, Xiao R. Huang, and Hui Y. Lan. 2012. "Smad3 Mediates ANG II-Induced Hypertensive Kidney Disease in Mice." *American Journal of Physiology. Renal Physiology* 302 (8). https://doi.org/10.1152/AJPRENAL.00595.2011.

Lorenzo, Andrea F.Gil, Valeria V. Costantino, Martin López Appiolaza, Valeria Cacciamani, Maria E. Benardon, Victoria Bocanegra, and Patricia G. Vallés. 2015. "Heat Shock Protein 70 and CHIP Promote Nox4 Ubiquitination and Degradation within the Losartan Antioxidative Effect in Proximal Tubule Cells." *Cellular Physiology and Biochemistry: International Journal of Experimental Cellular Physiology, Biochemistry, and Pharmacology* 36 (6): 2183–97. https://doi.org/10.1159/000430184.

Lorz, Corina, Alberto Ortiz, Pilar Justo, Silvia González-Cuadrado, Natalia Duque, Carmen Gómez-Guerrero, and Jesús Egido. 2000. "Proapoptotic Fas Ligand Is Expressed by Normal Kidney Tubular Epithelium and Injured Glomeruli." *Journal of the American Society of Nephrology: JASN* 11 (7): 1266–77. https://doi.org/10.1681/ASN.V1171266.

Lovisa, Sara, Valerie S. LeBleu, Björn Tampe, Hikaru Sugimoto, Komal Vadnagara, Julienne L. Carstens, Chia Chin Wu, Yohannes Hagos, Birgitta C Burckhardt, Tsvetelina Pentcheva-Hoang, Hersharan Nischal, James P Allison, Michael Zeisberg, Raghu Kalluri. 2015. "Epithelial-to-Mesenchymal Transition Induces Cell Cycle

Arrest and Parenchymal Damage in Renal Fibrosis." *Nature Medicine* 21 (9): 998–1009. https://doi.org/10.1038/NM.3902.

Maarouf, Omar H., Anusha Aravamudhan, Deepika Rangarajan, Tetsuro Kusaba, Victor Zhang, Jeremy Welborn, Daniel Gauvin, Xiuyun Hou, Rafael Kramann, and Benjamin D. Humphreys. 2016. "Paracrine Wnt1 Drives Interstitial Fibrosis without Inflammation by Tubulointerstitial Cross-Talk." *Journal of the American Society of Nephrology: JASN* 27 (3): 781–90. https://doi.org/10.1681/ASN.2014121188.

Macconi, Daniela, Mauro Abbate, Marina Morigi, Stefania Angioletti, Marilena Mister, Simona Buelli, Maria Bonomelli, Peter Mundel, Karlhans Endlich, Andrea Remuzzi, Giuseppe Remuzzi. 2006. "Permselective Dysfunction of Podocyte-Podocyte Contact upon Angiotensin II Unravels the Molecular Target for Renoprotective Intervention." *The American Journal of Pathology* 168 (4): 1073–85. https://doi.org/10.2353/AJPATH.2006.050701.

Macconi, Daniela, Chiara Chiabrando, Silvia Schiarea, Sistiana Aiello, Linda Cassis, Elena Gagliardini, Marina Noris, Simona Buelli, Carla Zoja, Daniela Corna, Caterina Mele, Roberto Fanelli, Giuseppe Remuzzi, Ariela Benigni. 2009. "Proteasomal Processing of Albumin by Renal Dendritic Cells Generates Antigenic Peptides." *Journal of the American Society of Nephrology: JASN* 20 (1): 123–30. https://doi.org/10.1681/ASN.2007111233.

Macconi, Daniela, Giuseppe Remuzzi, and Ariela Benigni. 2014. "Key Fibrogenic Mediators: Old Players. Renin-Angiotensin System." *Kidney International Supplements* 4 (1): 58–64. https://doi.org/10.1038/KISUP.2014.11.

Machida, Yuichi, Koichiro Kitamoto, Yasukatsu Izumi, Masayuki Shiota, Junji Uchida, Yukimi Kira, Tatsuya Nakatani, and Katsuyuki Miura. 2010. "Renal Fibrosis in Murine Obstructive Nephropathy Is Attenuated by Depletion of Monocyte Lineage, Not Dendritic Cells." *Journal of Pharmacological Sciences* 114 (4): 464–73. https://doi.org/10.1254/JPHS.10246FP.

Mack, Matthias. 2018. "Inflammation and Fibrosis." *Matrix Biology: Journal of the International Society for Matrix Biology* 68–69 (August): 106–21. https://doi.org/10.1016/J.MATBIO.2017.11.010.

Mackensen-Haen, S, R Bader, K E Grund, and A Bohle. 1981. "Correlations between Renal Cortical Interstitial Fibrosis, Atrophy of the Proximal Tubules and Impairment of the Glomerular Filtration Rate." *Clin Nephrol.* 1981. https://pubmed.ncbi.nlm.nih.gov/7237863/.

Mayer, Matthias P. 2018. "Intra-Molecular Pathways of Allosteric Control in Hsp70s." *Philosophical Transactions of the Royal Society of London. Series B, Biological Sciences* 373 (1749). https://doi.org/10.1098/RSTB.2017.0183.

McDonough, Holly, and Cam Patterson. 2003. "CHIP: A Link between the Chaperone and Proteasome Systems." *Cell Stress Chaperones.* 2003. https://pubmed.ncbi.nlm.nih.gov/15115282/.

Mendelsohn, F A. 1982. "Angiotensin II: Evidence for Its Role as an Intrarenal Hormone." *Kidney Int Suppl.* 1982. https://pubmed.ncbi.nlm.nih.gov/6182339/.

Meng, Xiao Ming, Xiao Ru Huang, Arthur C. K. Chung, Wei Qin, Xinli Shao, Peter Igarashi, Wenjun Ju, Erwin P. Bottinger, and Hui Yao Lan. 2010. "Smad2 Protects against TGF-Beta/Smad3-Mediated Renal Fibrosis." *Journal of the American Society of Nephrology: JASN* 21 (9): 1477–87. https://doi.org/10.1681/ASN.2009121244.

Mezzano, S. A., M. Ruiz-Ortega, and J. Egido. 2001. "Angiotensin II and Renal Fibrosis." *Hypertension* (Dallas, Tex.: 1979) 38 (3 Pt 2): 635–38. https://doi.org/10.1161/HY09T1.094234.

Mezzano, Sergio A., Miguel Barria, M. Alejandra Droguett, M. Eugenia Burgos, Leopoldo G. Ardiles, Claudio Flores, and Jesús Egido. 2001. "Tubular NF-KappaB and AP-1 Activation in Human Proteinuric Renal Disease." *Kidney International* 60 (4): 1366–77. https://doi.org/10.1046/J.1523-1755.2001.00941.X.

Mezzano, Sergio A., M. Alejandra Droguett, M. Eugenia Burgos, Leopoldo G. Ardiles, Claudio A. Aros, Italo Caorsi, and Jesús Egido. 2000. "Overexpression of Chemokines, Fibrogenic Cytokines, and Myofibroblasts in Human Membranous Nephropathy." *Kidney International* 57 (1): 147–58. https://doi.org/10.1046/J.1523-1755.2000.00830.X.

Mkaddem, Sanae Ben, Catherine Werts, Jean Michel Goujon, Marcelle Bens, Eric Pedruzzi, Eric Ogier-Denis, and Alain Vandewalle. 2009. "Heat Shock Protein Gp96 Interacts with Protein Phosphatase 5 and Controls Toll-like Receptor 2 (TLR2)-Mediated Activation of Extracellular Signal-Regulated Kinase (ERK) 1/2 in Post-Hypoxic Kidney Cells." *The Journal of Biological Chemistry* 284 (18): 12541–49. https://doi.org/10.1074/JBC.M808376200.

Mogk, Axel, Bernd Bukau, and Harm H. Kampinga. 2018. "Cellular Handling of Protein Aggregates by Disaggregation Machines." *Molecular Cell* 69 (2): 214–26. https://doi.org/10.1016/J.MOLCEL.2018.01.004.

Molitoris, B. A. 1991. "Ischemia-Induced Loss of Epithelial Polarity: Potential Role of the Actin Cytoskeleton." *The American Journal of Physiology* 260 (6 Pt 2). https://doi.org/10.1152/AJPRENAL.1991.260.6.F769.

Müller, Eva, Wolfgang Neuhofer, Anke Burger-Kentischer, Akihiro Ohno, Klaus Thurau, and Franz X. Beck. 1998. "Effects of Long-Term Changes in Medullary Osmolality on Heat Shock Proteins HSp25, HSP60, HSP72 and HSP73 in the Rat Kidney." *Pflugers Archiv : European Journal of Physiology* 435 (5): 705–12. https://doi.org/10.1007/S004240050572.

Müller, Eva, Wolfgang Neuhofer, Akihiro Ohno, Silvia Rucker, Klaus Thurau, and Franz X. Beck. 1996. "Heat Shock Proteins HSP25, HSP60, HSP72, HSP73 in Isoosmotic Cortex and Hyperosmotic Medulla of Rat Kidney." *Pflugers Archiv : European Journal of Physiology* 431 (4): 608–17. https://doi.org/10.1007/BF02191910.

Nangaku, Masaomi. 2006. "Chronic Hypoxia and Tubulointerstitial Injury: A Final Common Pathway to End-Stage Renal Failure." *Journal of the American Society of Nephrology: JASN* 17 (1): 17–25. https://doi.org/10.1681/ASN.2005070757.

Nangaku, Masaomi, and Kai Uwe Eckardt. 2007. "Hypoxia and the HIF System in Kidney Disease." *Journal of Molecular Medicine (Berlin, Germany)* 85 (12): 1325–30. https://doi.org/10.1007/S00109-007-0278-Y.

Nath, Karl A. 1992. "Tubulointerstitial Changes as a Major Determinant in the Progression of Renal Damage." *American Journal of Kidney Diseases: The Official Journal of the National Kidney Foundation* 20 (1): 1–17. https://doi.org/10.1016/S0272-6386(12)80312-X.

Navar, L. Gabriel, Lisa M. Harrison-Bernard, Akira Nishiyama, and Hiroyuki Kobori. 2002. "Regulation of Intrarenal Angiotensin II in Hypertension." *Hypertension (Dallas, Tex.: 1979)* 39 (2 Pt 2): 316–22. https://doi.org/10.1161/HY0202.103821.

Neuhofer, W., K. Lugmayr, M. L. Fraek, and F. X. Beck. 2001. "Regulated Overexpression of Heat Shock Protein 72 Protects Madin-Darby Canine Kidney Cells from the Detrimental Effects of High Urea Concentrations." *Journal of the American Society of Nephrology: JASN* 12 (12): 2565–71. https://doi.org/10.1681/ASN.V12122565.

Ni, Jun, Fuye Yang, Xiao Ru Huang, Jinxiu Meng, Jiaoyi Chen, Michael Bader, Josef M. Penninger, Erik Fung, Xue Qing Yu, and Hui Yao Lan. 2020. "Dual Deficiency of Angiotensin-Converting Enzyme-2 and Mas Receptor Enhances Angiotensin II-Induced Hypertension and Hypertensive Nephropathy." *Journal of Cellular and Molecular Medicine* 24 (22): 13093–103. https://doi.org/10.1111/JCMM.15914.

Nieto, M. Angela, Ruby Yun Y. J. Huang, Rebecca A A. Jackson, and Jean Paul P. Thiery. 2016. "EMT: 2016." *Cell* 166 (1): 21–45. https://doi.org/10.1016/J.CELL.2016.06.028.

Nillegoda, Nadinath B., Anne S. Wentink, and Bernd Bukau. 2018. "Protein Disaggregation in Multicellular Organisms." *Trends in Biochemical Sciences* 43 (4): 285–300. https://doi.org/10.1016/J.TIBS.2018.02.003.

Ninichuk, V., O. Gross, S. Segerer, R. Hoffmann, E. Radomska, A. Buchstaller, R. Huss, N. Akis, D. Schlöndorff, and H. J. Anders. 2006. "Multipotent Mesenchymal Stem Cells Reduce Interstitial Fibrosis but Do Not Delay Progression of Chronic Kidney Disease in Collagen4A3-Deficient Mice." *Kidney International* 70 (1): 121–29. https://doi.org/10.1038/SJ.KI.5001521.

Odunuga, Odutayo O., Judith A. Hornby, Christiane Bies, Richard Zimmermann, David J. Pugh, and Gregory L. Blatch. 2003. "Tetratricopeptide Repeat Motif-Mediated Hsc70-MSTI1 Interaction. Molecular Characterization of the Critical Contacts for Successful Binding and Specificity." *The Journal of Biological Chemistry* 278 (9): 6896–6904. https://doi.org/10.1074/JBC.M206867200.

Okada, Hirokazu, Tomohiro Kikuta, Tatsuya Kobayashi, Tsutomu Inoue, Yoshihiko Kanno, Masaharu Takigawa, Takeshi Sugaya, Jeffrey B. Kopp, and Hiromichi Suzuki. 2005. "Connective Tissue Growth Factor Expressed in Tubular Epithelium Plays a Pivotal Role in Renal Fibrogenesis." *Journal of the American Society of Nephrology: JASN* 16 (1): 133–43. https://doi.org/10.1681/ASN.2004040339.

O'Neill, Stephen, Ewen M. Harrison, James A. Ross, Stephen J. Wigmore, and Jeremy Hughes. 2014. "Heat-Shock Proteins and Acute Ischaemic Kidney Injury." *Nephron. Experimental Nephrology* 126 (4): 167–74. https://doi.org/10.1159/000363323.

Ortiz, Alberto, Corina Lorz, and Jesús Egido. 1999. "The Fas Ligand/Fas System in Renal Injury." *Nephrology, Dialysis, Transplantation: Official Publication of the European Dialysis and Transplant Association - European Renal Association* 14 (8): 1831–34. https://doi.org/10.1093/NDT/14.8.1831.

Patten, David A., Véronique N. Lafleur, Geneviève A. Robitaille, Denise A. Chan, Amato J. Giaccia, and Darren E. Richard. 2010. "Hypoxia-Inducible Factor-1 Activation in Nonhypoxic Conditions: The Essential Role of Mitochondrial-Derived Reactive Oxygen Species." *Molecular Biology of the Cell* 21 (18): 3247–57. https://doi.org/10.1091/MBC.E10-01-0025.

Pockley, A. Graham. 2003. "Heat Shock Proteins as Regulators of the Immune Response." *Lancet* (London, England) 362 (9382): 469–76. https://doi.org/10.1016/S0140-6736(03)14075-5.

Polichnowski, Aaron J., Rongpei Lan, Hui Geng, Karen A. Griffin, Manjeri A. Venkatachalam, and Anil K. Bidani. 2014. "Severe Renal Mass Reduction Impairs Recovery and Promotes Fibrosis after AKI." *Journal of the American Society of Nephrology* 25 (7): 1496–1507. https://doi.org/10.1681/ASN.2013040359/-/DCSUPPLEMENTAL.

Prozialeck, Walter C., and Joshua R. Edwards. 2007. "Cell Adhesion Molecules in Chemically-Induced Renal Injury." *Pharmacology & Therapeutics* 114 (1): 74–93. https://doi.org/10.1016/J.PHARMTHERA.2007.01.001.

Qu, Baoge, Yiguo Jia, Yuanxun Liu, Hui Wang, Guangying Ren, and Hong Wang. 2015. "The Detection and Role of Heat Shock Protein 70 in Various Nondisease Conditions and Disease Conditions: A Literature Review." *Cell Stress & Chaperones* 20 (6): 885–92. https://doi.org/10.1007/S12192-015-0618-8.

Risdon, R. A., J. C. Sloper, and H. E. De Wardener. 1968. "Relationship between Renal Function and Histological Changes Found in Renal-Biopsy Specimens from Patients with Persistent Glomerular Nephritis." *Lancet (London, England)* 2 (7564): 363–66. https://doi.org/10.1016/S0140-6736(68)90589-8.

Rodrigues-Díez, Raquel, Gisselle Carvajal-González, Elsa Sánchez-López, Juan Rodríguez-Vita, Raúl Rodrigues Díez, Rafael Selgas, Alberto Ortiz, Jesús Egido, Sergio Mezzano, and Marta Ruiz-Ortega. 2008. "Pharmacological Modulation of Epithelial Mesenchymal Transition Caused by Angiotensin II. Role of ROCK and MAPK Pathways." *Pharmaceutical Research* 25 (10): 2447–61. https://doi.org/10.1007/S11095-008-9636-X.

Rodrigues-Diez, Raúl R., Ana Belen Garcia-Redondo, Macarena Orejudo, Raquel Rodrigues-Diez, Ana Maria Briones, Enrique Bosch-Panadero, Gyorgy Kery, Janos Pato, Alberto Ortiz, Mercedes Salaices, Jesus Egido, Marta Ruiz-Ortega. 2015. "The C-Terminal Module IV of Connective Tissue Growth Factor, through EGFR/Nox1 Signaling, Activates the NF-KB Pathway and Proinflammatory Factors in Vascular Smooth Muscle Cells." *Antioxidants & Redox Signaling* 22 (1): 29–47. https://doi.org/10.1089/ARS.2013.5500.

Rodríguez-Iturbe, Bernardo, and Guillermo García García. 2010. "The Role of Tubulointerstitial Inflammation in the Progression of Chronic Renal Failure." *Nephron. Clinical Practice* 116 (2). https://doi.org/10.1159/000314656.

Roh, Jong Seong, and Dong Hyun Sohn. 2018. "Damage-Associated Molecular Patterns in Inflammatory Diseases." *Immune Network* 18 (4). https://doi.org/10.4110/IN.2018.18.E27.

Rosenzweig, Rina, Nadinath B. Nillegoda, Matthias P. Mayer, and Bernd Bukau. 2019. "The Hsp70 Chaperone Network." *Nature Reviews. Molecular Cell Biology* 20 (11): 665–80. https://doi.org/10.1038/S41580-019-0133-3.

Ruete, María Celeste, Liliana C. Carrizo, and Patricia G. Vallés. 2008. "Na+/K+ -ATPase Stabilization by Hsp70 in the Outer Stripe of the Outer Medulla in Rats during Recovery from a Low-Protein Diet." *Cell Stress & Chaperones* 13 (2): 157–67. https://doi.org/10.1007/S12192-008-0021-9.

Sampangi, Sandeep, Xiangju Wang, Kenneth W. Beagley, Travis Klein, Sadia Afrin, Helen Healy, Ray Wilkinson, and Andrew J. Kassianos. 2015. "Human Proximal Tubule Epithelial Cells Modulate Autologous B-Cell Function." *Nephrology, Dialysis, Transplantation: Official Publication of the European Dialysis and Transplant Association - European Renal Association* 30 (10): 1674–83. https://doi.org/10.1093/NDT/GFV242.

Sánchez-López, Elsa, Sandra Rayego, Raquel Rodrigues-Díez, Javier Sánchez Rodriguez, Raúl Rodrigues-Díez, Juan Rodríguez-Vita, Gisselle Carvajal, Luiz Stark Aroeira, Rafael Selgas, Sergio A Mezzano, Alberto Ortiz, Jesús Egido, Marta Ruiz-Ortega. 2009. "CTGF Promotes Inflammatory Cell Infiltration of the Renal Interstitium by Activating NF-KappaB." *Journal of the American Society of Nephrology: JASN* 20 (7): 1513–26. https://doi.org/10.1681/ASN.2008090999.

Sanz, Ana B., M. Concepcion Izquierdo, Maria Dolores Sanchez-Niño, Alvaro C. Ucero, Jesus Egido, Marta Ruiz-Ortega, Adrián Mario Ramos, Chaim Putterman, and Alberto Ortiz. 2014. "TWEAK and the Progression of Renal Disease: Clinical Translation." *Nephrology, Dialysis, Transplantation: Official Publication of the European Dialysis and Transplant Association - European Renal Association* 29 Suppl 1 (Suppl 1). https://doi.org/10.1093/NDT/GFT342.

Sanz, Ana Belen, Maria D. Sanchez-Nĩo, and Alberto Ortiz. 2011. "TWEAK, a Multifunctional Cytokine in Kidney Injury." *Kidney International* 80 (7): 708–18. https://doi.org/10.1038/KI.2011.180.

Scheufler, Clemens, Achim Brinker, Gleb Bourenkov, Stefano Pegoraro, Luis Moroder, Hans Bartunik, F. Ulrich Hartl, and Ismail Moarefi. 2000. "Structure of TPR Domain-Peptide Complexes: Critical Elements in the Assembly of the Hsp70-Hsp90 Multichaperone Machine." *Cell* 101 (2): 199–210. https://doi.org/10.1016/S0092-8674(00)80830-2.

Schlecht, Rainer, Annette H. Erbse, Bernd Bukau, and Matthias P. Mayer. 2011. "Mechanics of Hsp70 Chaperones Enables Differential Interaction with Client Proteins." *Nature Structural & Molecular Biology* 18 (3): 345–51. https://doi.org/10.1038/NSMB.2006.

Schnaper, H. William. 2017. "The Tubulointerstitial Pathophysiology of Progressive Kidney Disease." *Advances in Chronic Kidney Disease* 24 (2): 107–16. https://doi.org/10.1053/J.ACKD.2016.11.011.

Schober, A., E. Müller, K. Thurau, and F. X. Beck. 1997. "The Response of Heat Shock Proteins 25 and 72 to Ischaemia in Different Kidney Zones." *Pflugers Archiv : European Journal of Physiology* 434 (3): 292–99. https://doi.org/10.1007/S004240050399.

Segerer, Stephan, and Charles E. Alpers. 2003. "Chemokines and Chemokine Receptors in Renal Pathology." *Current Opinion in Nephrology and Hypertension* 12 (3): 243–49. https://doi.org/10.1097/00041552-200305000-00004.

Seikaly, M. G., B. S. Arant, and F. D. Seney. 1990. "Endogenous Angiotensin Concentrations in Specific Intrarenal Fluid Compartments of the Rat." *The Journal of Clinical Investigation* 86 (4): 1352–57. https://doi.org/10.1172/JCI114846.

Sleeman, Jonathan P., and Jean Paul Thiery. 2011. "SnapShot: The Epithelial-Mesenchymal Transition." *Cell* 145 (1). https://doi.org/10.1016/J.CELL.2011.03.029.

Sone, M., G. J. Albrecht, A. Dorge, K. Thurau, and F. X. Beck. 1993. "Osmotic Adaptation of Renal Medullary Cells during Transition from Chronic Diuresis to Antidiuresis." *The American Journal of Physiology* 264 (4 Pt 2). https://doi.org/10.1152/AJPRENAL.1993.264.4.F722.

Stankiewicz, Marta, Rainer Nikolay, Vladimir Rybin, and Matthias P. Mayer. 2010. "CHIP Participates in Protein Triage Decisions by Preferentially Ubiquitinating Hsp70-Bound Substrates." *The FEBS Journal* 277 (16): 3353–67. https://doi.org/10.1111/J.1742-4658.2010.07737.X.

Sun, Shiren, Xiaoxuan Ning, Yanqi Zhang, Yuanyuan Lu, Yongzhan Nie, Shuang Han, Lili Liu, Rui Du, Lin Xia, Lijie He, Daiming Fan. 2009. "Hypoxia-Inducible Factor-1alpha Induces Twist Expression in Tubular Epithelial Cells Subjected to Hypoxia, Leading to Epithelial-to-Mesenchymal Transition." *Kidney International* 75 (12): 1278–87. https://doi.org/10.1038/KI.2009.62.

Sun, Yu Bo Yang, Xinli Qu, Georgina Caruana, and Jinhua Li. 2016. "The Origin of Renal Fibroblasts/Myofibroblasts and the Signals That Trigger Fibrosis." *Differentiation; Research in Biological Diversity* 92 (3): 102–7. https://doi.org/10.1016/J.DIFF.2016.05.008.

Szeto, Stephen G., Masahiro Narimatsu, Mingliang Lu, Xiaolin He, Ahmad M. Sidiqi, Monica F. Tolosa, Lauren Chan, Krystale De Freitas, Janne Folke Bialik, Syamantak Majumder, Stellar Boo, Boris Hinz, Qinghong Dan, Andrew Advani, Rohan John, Jeffrey L Wrana, Andras Kapus, Darren A Yuen. 2016. "YAP/TAZ Are Mechanoregulators of TGF- β-Smad Signaling and Renal Fibrogenesis." *Journal of the American Society of Nephrology: JASN* 27 (10): 3117–28. https://doi.org/10.1681/ASN.2015050499.

Tan, Ruoyun, Weichun He, Xia Lin, Lawrence P. Kiss, and Youhua Liu. 2008. "Smad Ubiquitination Regulatory Factor-2 in the Fibrotic Kidney: Regulation, Target Specificity, and Functional Implication." *American Journal of Physiology. Renal Physiology* 294 (5). https://doi.org/10.1152/AJPRENAL.00323.2007.

Tang, Lin, Run Yi, Bing Yang, Hui Li, Huijuan Chen, and Zhangsuo Liu. 2012. "Valsartan Inhibited HIF-1α Pathway and Attenuated Renal Interstitial Fibrosis in Streptozotocin-Diabetic Rats." *Diabetes Research and Clinical Practice* 97 (1): 125–31. https://doi.org/10.1016/J.DIABRES.2012.01.037.

Tesch, G. H. 2008. "MCP-1/CCL2: A New Diagnostic Marker and Therapeutic Target for Progressive Renal Injury in Diabetic Nephropathy." *American Journal of Physiology. Renal Physiology* 294 (4). https://doi.org/10.1152/AJPRENAL.00016.2008.

Ushio-Fukai, Masuko, and R. Wayne Alexander. 2006. "Caveolin-Dependent Angiotensin II Type 1 Receptor Signaling in Vascular Smooth Muscle." *Hypertension (Dallas, Tex.: 1979)* 48 (5): 797–803. https://doi.org/10.1161/01.HYP.0000242907.70697.5D.

Vanderwaal, Robert P., Leonard B. Maggi, Jason D. Weber, Clayton R. Hunt, and Joseph L. Roti Roti. 2009. "Nucleophosmin Redistribution Following Heat Shock: A Role in Heat-Induced Radiosensitization." *Cancer Research* 69 (16): 6454–62. https://doi.org/10.1158/0008-5472.CAN-08-4896.

Venkatachalam, Manjeri A., Joel M. Weinberg, Wilhelm Kriz, and Anil K. Bidani. 2015. "Failed Tubule Recovery, AKI-CKD Transition, and Kidney Disease Progression." *Journal of the American Society of Nephrology: JASN* 26 (8): 1765–76. https://doi.org/10.1681/ASN.2015010006.

Verrecchia, Franck, Mon Li Chu, and Alain Mauviel. 2001. "Identification of Novel TGF-Beta/Smad Gene Targets in Dermal Fibroblasts Using a Combined CDNA Microarray/Promoter Transactivation Approach." *The Journal of Biological Chemistry* 276 (20): 17058–62. https://doi.org/10.1074/JBC.M100754200.

Vogel, Markus, Matthias P. Mayer, and Bernd Bukau. 2006. "Allosteric Regulation of Hsp70 Chaperones Involves a Conserved Interdomain Linker." *The Journal of Biological Chemistry* 281 (50): 38705–11. https://doi.org/10.1074/JBC.M609020200.

Walther, Dirk M., Prasad Kasturi, Min Zheng, Stefan Pinkert, Giulia Vecchi, Prajwal Ciryam, Richard I. Morimoto, Christopher M Dobson, Michele Vendruscolo, Matthias Mann, F Ulrich Hartl. 2015. "Widespread Proteome Remodeling and Aggregation in Aging C. Elegans." *Cell* 161 (4): 919–32. https://doi.org/10.1016/J.CELL.2015.03.032.

Wang, Jing, Yi Wen, Lin Li Lv, Hong Liu, Ri Ning Tang, Kun Ling Ma, and Bi Cheng Liu. 2015. "Involvement of Endoplasmic Reticulum Stress in Angiotensin II-Induced NLRP3 Inflammasome Activation in Human Renal Proximal Tubular Cells in Vitro." *Acta Pharmacologica Sinica* 36 (7): 821–30. https://doi.org/10.1038/APS.2015.21.

Wang, S., H. Diao, Q. Guan, W. W. Cruikshank, T. L. Delovitch, A. M. Jevnikar, and C. Du. 2008. "Decreased Renal Ischemia-Reperfusion Injury by IL-16 Inactivation." *Kidney International* 73 (3): 318–26. https://doi.org/10.1038/SJ.KI.5002692.

Wang, Wansheng, Xiao R. Huang, Ellery Canlas, Kazuhiro Oka, Luan D. Truong, Chuxia Deng, Neil A. Bhowmick, Wenjun Ju, Erwin P. Bottinger, and Hui Y. Lan. 2006. "Essential Role of Smad3 in Angiotensin II-Induced Vascular Fibrosis." *Circulation Research* 98 (8): 1032–39. https://doi.org/10.1161/01.RES.0000218782.52610.DC.

Wang, Yinqiu, Jian Chang, Bing Yao, Aolei Niu, Emily Kelly, Matthew C. Breeggemann, Sherry L. Abboud Werner, Raymond C. Harris, and Ming Zhi Zhang. 2015. "Proximal Tubule-Derived Colony Stimulating Factor-1 Mediates Polarization of Renal Macrophages and Dendritic Cells, and Recovery in Acute Kidney Injury." *Kidney International* 88 (6): 1274–82. https://doi.org/10.1038/KI.2015.295.

Wang, Yiping, Juchuan Chen, Liguang Chen, Yuet Ching Tay, Gopala K. Rangan, and David C. H. Harris. 1997. "Induction of Monocyte Chemoattractant Protein-1 in Proximal Tubule Cells by Urinary Protein." *Journal of the American Society of Nephrology: JASN* 8 (10): 1537–45. https://doi.org/10.1681/ASN.V8101537.

Wang, Yiping, Gopala K. Rangan, Bryan Goodwin, Yuet Ching Tay, Yang Wang, and David C. H. Harris. 2000. "Lipopolysaccharide-Induced MCP-1 Gene Expression in Rat Tubular Epithelial Cells Is Nuclear Factor-KappaB Dependent." *Kidney International* 57 (5): 2011–22. https://doi.org/10.1046/J.1523-1755.2000.00051.X.

Wang, Zhengchao, Lin Tang, Qing Zhu, Fan Yi, Fan Zhang, Pin Lan Li, and Ningjun Li. 2011. "Hypoxia-Inducible Factor-1α Contributes to the Profibrotic Action of

Angiotensin II in Renal Medullary Interstitial Cells." *Kidney International* 79 (3): 300–310. https://doi.org/10.1038/KI.2010.326.

Wang, Zhiyong, Jonathan M. Gall, Ramon G. B. Bonegio, Andrea Havasi, Clayton R. Hunt, Michael Y. Sherman, John H. Schwartz, and Steven C. Borkan. 2011. "Induction of Heat Shock Protein 70 Inhibits Ischemic Renal Injury." *Kidney International* 79 (8): 861–70. https://doi.org/10.1038/KI.2010.527.

Wang, Zhiyong, Andrea Havasi, Aaron A. Beeler, and Steven C. Borkan. 2022. "Mechanisms of Nucleophosmin (NPM)-Mediated Regulated Cell Death Elucidated by Hsp70 during Renal Ischemia." *Apoptosis: An International Journal on Programmed Cell Death* 27 (1–2): 22–33. https://doi.org/10.1007/S10495-021-01696-8.

Wang, Zhiyong, Erdjan Salih, Chinaemere Igwebuike, Ryan Mulhern, Ramon G. Bonegio, Andrea Havasi, and Steven C. Borkan. 2019. "Nucleophosmin Phosphorylation as a Diagnostic and Therapeutic Target for Ischemic AKI." *Journal of the American Society of Nephrology: JASN* 30 (1): 50–62. https://doi.org/10.1681/ASN.2018040401.

Wei, Xiaofan, Yang Xia, Feng Li, Yan Tang, Jing Nie, Youhua Liu, Zhanmei Zhou, Hongquan Zhang, and Fan Fan Hou. 2013. "Kindlin-2 Mediates Activation of TGF-β/Smad Signaling and Renal Fibrosis." *Journal of the American Society of Nephrology: JASN* 24 (9): 1387–98. https://doi.org/10.1681/ASN.2012101041.

Weigert, C., K. Brodbeck, K. Klopfer, H. Häring, and E. Schleicher. 2002. "Angiotensin II Induces Human TGF-Beta 1 Promoter Activation: Similarity to Hyperglycaemia." *Diabetologia* 45 (6): 890–98. https://doi.org/10.1007/S00125-002-0843-4.

Weiler, M., L. Kachko, C. Chaimovitz, C. Van Kooten, and A. Douvdevani. 2001. "CD40 Ligation Enhances IL-15 Production by Tubular Epithelial Cells." *Journal of the American Society of Nephrology: JASN* 12 (1): 80–87. https://doi.org/10.1681/ASN.V12.1.80.

Weinberg, Joel M., and Bruce A. Molitoris. 2009. "Illuminating Mitochondrial Function and Dysfunction Using Multiphoton Technology." *Journal of the American Society of Nephrology: JASN* 20 (6): 1164–66. https://doi.org/10.1681/ASN.2009040419.

Weisheit, Christina K., Daniel R. Engel, and Christian Kurts. 2015. "Dendritic Cells and Macrophages: Sentinels in the Kidney." *Clinical Journal of the American Society of Nephrology: CJASN* 10 (10): 1841–51. https://doi.org/10.2215/CJN.07100714.

Welch, W. J. 1993. "How Cells Respond to Stress." *Scientific American* 268 (5): 56–64. https://doi.org/10.1038/SCIENTIFICAMERICAN0593-56.

Wen, Yi, Yiran Liu, Taotao Tang, Linli Lv, Hong Liu, Kunling Ma, and Bicheng Liu. 2016. "NLRP3 Inflammasome Activation Is Involved in Ang II-Induced Kidney Damage via Mitochondrial Dysfunction." *Oncotarget* 7 (34): 54290–302. https://doi.org/10.18632/ONCOTARGET.11091.

Why, S. K. Van, F. Hildebrandt, T. Ardito, A. S. Mann, N. J. Siegel, and M. Kashgarian. 1992. "Induction and Intracellular Localization of HSP-72 after Renal Ischemia." *The American Journal of Physiology* 263 (5 Pt 2). https://doi.org/10.1152/AJPRENAL.1992.263.5.F769.

Wickner, Sue, Michael R. Maurizi, and Susan Gottesman. 1999. "Posttranslational Quality Control: Folding, Refolding, and Degrading Proteins." *Science* (New York, N.Y.) 286 (5446): 1888–93. https://doi.org/10.1126/SCIENCE.286.5446.1888.

Wolf, G., G. Zahner, R. Schroeder, and R. A. K. Stahl. 1996. "Transforming Growth Factor Beta Mediates the Angiotensin-II-Induced Stimulation of Collagen Type IV Synthesis in Cultured Murine Proximal Tubular Cells." *Nephrology, Dialysis, Transplantation: Official Publication of the European Dialysis and Transplant Association - European Renal Association* 11 (2): 263–69. https://doi.org/10.1093/OXFORDJOURNALS.NDT.A027251.

Wolfs, Tim G. A. M., Wim A. Buurman, Annemarie van Schadewijk, Bart de Vries, Marc A. R. C. Daemen, Pieter S. Hiemstra, and Cornelis van 't Veer. 2002. "In Vivo Expression of Toll-like Receptor 2 and 4 by Renal Epithelial Cells: IFN-Gamma and TNF-Alpha Mediated up-Regulation during Inflammation." *Journal of Immunology* (Baltimore, Md.: 1950) 168 (3): 1286–93. https://doi.org/10.4049/JIMMUNOL.168.3.1286.

Wong, Wai Keong, Helen Robertson, Helen P. Carroll, Simi Ali, and John A. Kirby. 2003. "Tubulitis in Renal Allograft Rejection: Role of Transforming Growth Factor-Beta and Interleukin-15 in Development and Maintenance of CD103+ Intraepithelial T Cells." *Transplantation* 75 (4): 505–14. https://doi.org/10.1097/01.TP.0000045708.39879.C7.

Wu, Huiling, Gang Chen, Kate R. Wyburn, Jianlin Yin, Patrick Bertolino, Josette M. Eris, Stephen I. Alexander, Alexandra F. Sharland, and Steven J. Chadban. 2007. "TLR4 Activation Mediates Kidney Ischemia/Reperfusion Injury." *The Journal of Clinical Investigation* 117 (10): 2847–59. https://doi.org/10.1172/JCI31008.

Wu, Huiling, Jin Ma, Peng Wang, Theresa M. Corpuz, Usha Panchapakesan, Kate R. Wyburn, and Steven J. Chadban. 2010. "HMGB1 Contributes to Kidney Ischemia Reperfusion Injury." *Journal of the American Society of Nephrology: JASN* 21 (11): 1878–90. https://doi.org/10.1681/ASN.2009101048.

Wyse, Bruce D., Ian A. Prior, Hongwei Qian, Isabel C. Morrow, Susan Nixon, Cornelia Muncke, Teymuras V. Kurzchalia, Walter G. Thomas, Robert G. Parton, and John F. Hancock. 2003. "Caveolin Interacts with the Angiotensin II Type 1 Receptor during Exocytic Transport but Not at the Plasma Membrane." *The Journal of Biological Chemistry* 278 (26): 23738–46. https://doi.org/10.1074/JBC.M212892200.

Yang, De, Zhen Han, and Joost J. Oppenheim. 2017. "Alarmins and Immunity." *Immunological Reviews* 280 (1): 41–56. https://doi.org/10.1111/IMR.12577.

Yang, Fuye, Arthur C. K. Chung, Xiao Ru Huang, and Hui Yao Lan. 2009. "Angiotensin II Induces Connective Tissue Growth Factor and Collagen I Expression via Transforming Growth Factor-Beta-Dependent and -Independent Smad Pathways: The Role of Smad3." *Hypertension* (Dallas, Tex.: 1979) 54 (4): 877–84. https://doi.org/10.1161/HYPERTENSIONAHA.109.136531.

Yang, Li, Tatiana Y. Besschetnova, Craig R. Brooks, Jagesh V. Shah, and Joseph V. Bonventre. 2010. "Epithelial Cell Cycle Arrest in G2/M Mediates Kidney Fibrosis after Injury." *Nature Medicine* 16 (5): 535–43. https://doi.org/10.1038/NM.2144.

Yang, Yonghua, Warren Fiskus, Bao Yong, Peter Atadja, Yoshinori Takahashi, Tej K. Pandita, Hong Gang Wang, and Kapil N. Bhalla. 2013. "Acetylated Hsp70 and KAP1-

Mediated Vps34 SUMOylation Is Required for Autophagosome Creation in Autophagy." *Proceedings of the National Academy of Sciences of the United States of America* 110 (17): 6841–46. https://doi.org/10.1073/PNAS.1217692110.

Yang, Yunbo, Zhu Xu Zhang, Dameng Lian, Aaron Haig, Rabindra N. Bhattacharjee, and Anthony M. Jevnikar. 2015. "IL-37 Inhibits IL-18-Induced Tubular Epithelial Cell Expression of pro-Inflammatory Cytokines and Renal Ischemia-Reperfusion Injury." *Kidney International* 87 (2): 396–408. https://doi.org/10.1038/KI.2014.295.

Yard, B. A., M. R. Daha, M. Kooymans-Couthino, J. A. Bruijn, M. E. Paape, E. Schrama, L. A. Van Es, and F. J. Van Der Woude. 1992. "IL-1 Alpha Stimulated TNF Alpha Production by Cultured Human Proximal Tubular Epithelial Cells." *Kidney International* 42 (2): 383–89. https://doi.org/10.1038/KI.1992.299.

Zamyatina, Alla, and Holger Heine. 2020. "Lipopolysaccharide Recognition in the Crossroads of TLR4 and Caspase-4/11 Mediated Inflammatory Pathways." *Frontiers in Immunology* 11 (November). https://doi.org/10.3389/FIMMU.2020.585146.

Zhao, You Yang, Yang Liu, Radu Virgil Stan, Lian Fan, Yusu Gu, Nancy Dalton, Po Hsien Chu, Kirk Peterson, John Ross, and Kenneth R. Chien. 2002. "Defects in Caveolin-1 Cause Dilated Cardiomyopathy and Pulmonary Hypertension in Knockout Mice." *Proceedings of the National Academy of Sciences of the United States of America* 99 (17): 11375–80. https://doi.org/10.1073/PNAS.172360799.

Zhuo, Jia L., Oscar A. Carretero, and Xiao C. Li. 2006. "Effects of AT1 Receptor-Mediated Endocytosis of Extracellular Ang II on Activation of Nuclear Factor-Kappa B in Proximal Tubule Cells." *Annals of the New York Academy of Sciences* 1091: 336–45. https://doi.org/10.1196/ANNALS.1378.078.

Zhuo, Jia L., John D. Imig, Timothy G. Hammond, Sheyla Orengo, Edmund Benes, and L. Gabriel Navar. 2002. "Ang II Accumulation in Rat Renal Endosomes during Ang II-Induced Hypertension: Role of AT(1) Receptor." *Hypertension (Dallas, Tex. : 1979)* 39 (1): 116–21. https://doi.org/10.1161/HY0102.100780.

Zhuo, Jia L., and Xiao C. Li. 2011. "New Insights and Perspectives on Intrarenal Renin-Angiotensin System: Focus on Intracrine/Intracellular Angiotensin II." *Peptides* 32 (7): 1551–65. https://doi.org/10.1016/J.PEPTIDES.2011.05.012. 2013. "Proximal Nephron." *Comprehensive Physiology* 3 (3): 1079–1123. https://doi.org/10.1002/CPHY.C110061.

Zhuravleva, Anastasia, and Lila M. Gierasch. 2011. "Allosteric Signal Transmission in the Nucleotide-Binding Domain of 70-KDa Heat Shock Protein (Hsp70) Molecular Chaperones." *Proceedings of the National Academy of Sciences of the United States of America* 108 (17): 6987–92. https://doi.org/10.1073/PNAS.1014448108/-/DCSUPPLEMENTAL/PNAS.1014448108_SI.PDF.

Zuiderweg, Erik R. P., Lawrence E. Hightower, and Jason E. Gestwicki. 2017. "The Remarkable Multivalency of the Hsp70 Chaperones." *Cell Stress & Chaperones* 22 (2): 173–89. https://doi.org/10.1007/S12192-017-0776-Y

Chapter 3

Targeting Extracellular Polymeric Substances of Pathogenic Bacteria: *In Silico, In Vitro*, and Applied Strategies

Samaria Lisdeth Gutierrez-Pacheco[1]
Andres Alvarez-Armenta[2]
M. Melissa Gutierrez-Pacheco[3]
Luis Alberto Ortega-Ramírez[3]
and J. Fernando Ayala-Zavala[4,*]
[1]Departamento de Biotecnología y Ciencias Alimentarias, Instituto Tecnológico de Sonora, Ciudad Obregón, Sonora, México
[2]Departamento de Microbiología Molecular, Instituto de Biotecnología, Universidad Nacional Autónoma de México, Cuernavaca, Morelos, México
[3]Universidad Estatal de Sonora, San Luis Río Colorado, Sonora, México
[4]Coordinación de Tecnología de Alimentos de Origen Vegetal, Centro de Investigación en Alimentación y Desarrollo, A.C. Hermosillo, Sonora, México

Abstract

Pathogenic bacteria are known for their ability to form biofilms, a complex communities of bacterial cells embedded in extracellular polymeric substances (EPS). The EPS, consisting of proteins, polysaccharides, and DNA, acts as a protective barrier that shields the bacteria from antimicrobial agents, making biofilm eradication challenging. While cellulose plays a crucial role in biofilm formation, targeting other components, such as proteins, lipids, and DNA, is important to disrupt biofilm stability. In the search for effective

[*] Corresponding Author's E-mail: jayala@ciad.mx.

In: Advances in Biology. Volume 5
Editor: Charles D. Grant
ISBN: 979-8-89113-441-6

antibiofilm strategies, phytochemicals from plant extracts have emerged as promising candidates. Phytochemicals offer diverse bioactive compounds, including terpenes, coumarins, flavonoids, benzoic acid derivatives, tannins, and other phytochemical classes. Computational analysis using molecular docking and dynamics simulation can be employed to identify and evaluate the inhibitory effects of phytochemicals on target proteins related to EPS. These *in silico* techniques provide insights aiding in selecting potential candidates for further investigation. Following the computational analysis, *in vitro* experiments can be conducted to validate the inhibitory activity of the chosen phytochemicals on the target enzymes. These experiments involve exposing pathogenic bacteria to phytochemicals and measuring their effects on the synthesis of polymeric substances. Various assays, such as enzymatic activity assays and biofilm quantification assays, can be performed to assess the impact of phytochemicals on biofilm formation and stability. By integrating computational analysis, *in vitro* experiments, and applied strategies, novel antibiofilm approaches can be developed. The goal is to identify phytochemicals that effectively disrupt the synthesis of polymeric substances involved in biofilm formation and stability. This research can lead to the development of practical applications for combating biofilm-associated infections and improving antimicrobial strategies. In conclusion, this chapter will discuss using computational analysis, *in vitro* experiments, and applied methods to provide a comprehensive approach to target the synthesis of polymeric substances in pathogenic bacteria. This multidisciplinary approach holds promise for the development of effective antibiofilm interventions and the advancement of antimicrobial strategies.

Introduction

Infectious diseases initiated by pathogenic bacteria are a world health challenge that conduces to the search for innovative strategies for their prevention and treatment. The resistance mechanisms employed by bacteria for survival and virulence comprise the formation of biofilms. Biofilms are formed by a network of extracellular polymeric substances (EPS) conformed by a matrix of exopolysaccharides, proteins, lipids, and nucleic acids providing structural scaffold and protection to the bacteria. The variability in the composition of the EPS matrix and the interactions among the various components add new levels of complexity and provide challenges for developing EPS-targeting antimicrobials. Recent research has been increasingly focused on developing innovative approaches to target and disrupt bacterial biofilms. Phytochemicals can act as potential inhibitors of

EPS. Still, the inclusion of computational modeling to predict the interactions ligand-receptors and the evaluation of the stability of the interaction by molecular dynamics provides knowledge about the mechanism of action. Also, in silico techniques offer useful tools to correlate *in vitro* and in *vivo* behaviors to the phytochemical compounds or constituents of the plant extract used.

1. Extracellular Polymeric Substances (EPS) in Pathogenic Bacteria

In the natural environment, microorganisms are commonly found living on surfaces or closely associated with them, and they tend to exist predominantly as biofilms. Biofilms play an essential role in bacterial pathogenesis because, in this sessile mode of growth, bacteria became even 1000-fold more resistant to antimicrobial treatments than planktonic bacteria, causing recurrent and persistent infections as well as high morbidity and mortality (Hu et al., 2022). Biofilm is a consortium of bacterial cells embedded in a self-produced matrix composed of extracellular polymeric substances (EPS). It has been generalized that five steps conform biofilm formation process: 1) the initial reversible bacterial adherence to biotic or abiotic surfaces when cell density increases, 2) an irreversible adhesion helped by the EPS secretion, 3) microcolony formation by the aggregation of daughter cells, 4) the maturation resulted by the continuous secretion of EPS and cell multiplication, and 5) the dispersion of biofilm cells to colonize new surfaces (Ortega-Ramirez et al., 2021).

Biofilm matrix is a complex and dynamic structure, and EPS is a crucial element in establishing and maintaining the structure and properties of the biofilm. EPS synthesis and their spatial organization change between species, but in general, a diverse group of biomolecules can be grouped into two major categories: (i) molecules associated within the cell surface, such as type IV pili, flagella, and functional amyloids and (ii) those secreted out of the cell system such as proteins, exopolysaccharides, lipids, eDNA, and eRNA, which can be found in different proportions among bacteria (Ortega-Ramirez et al., 2021). EPS components play different but essential roles in biofilms, as observed in Figure 1 (Flemming et al., 2016). EPS's important role in biofilm structure and stability is helped by chemical interactions such as non-covalent bonding that involves weak physicochemical forces between microbial cells and EPS components, conferring cohesion and viscoelasticity to biofilm structure (Di Martino, 2018).

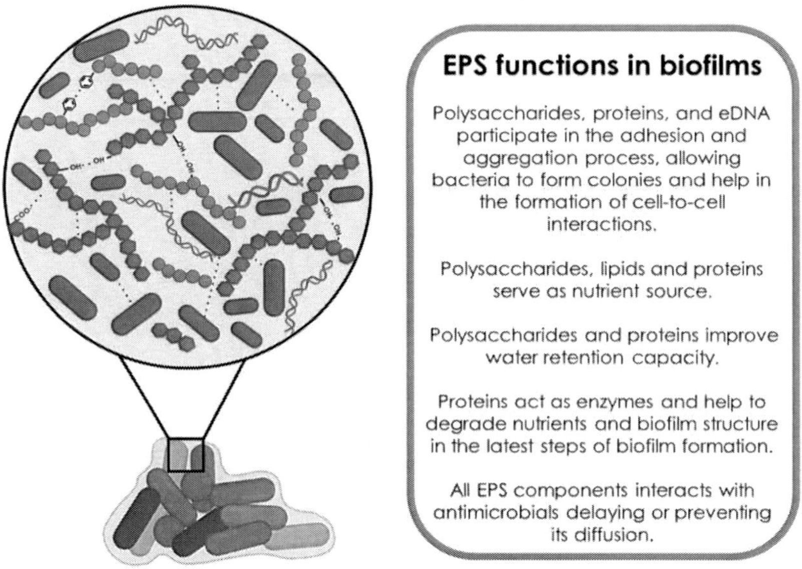

Figure 1. EPS components play vital functions in pathogenic bacteria biofilms.

One of the main functions of EPS is to prevent the penetration of dangerous molecules, such as antimicrobials, into biofilm cells, interacting with chemicals, making them non-bioavailable, and degrading/hydrolyzing hazardous compounds. This characteristic makes EPS potentially dangerous in food and medicine (Ortega-Ramirez et al., 2021). Developing new therapeutic strategies for persistent infections caused by human pathogenic bacteria requires knowing EPS composition and understanding how biofilm EPS protects against the immune system and antimicrobials (Gunn et al., 2016). Table 1 shows the main EPS of common human pathogenic bacteria reported in the literature.

It can be observed that bacteria secrete different types of EPS components; for example, *Pseudomonas aeruginosa* secretes three different exopolysaccharides, PsI, Pel, and alginate; each of them confers different characteristics to *P. aeruginosa* biofilms. The expression of alginate confers the mucoid phenotype, whereas Psl and Pel are expressed in non-mucoid strains (Colvin et al., 2012). On the other hand, in *Salmonella* species biofilms, polysaccharides include cellulose, colanic acid, O-antigen, Vi-antigen, and curli fimbriae. Specifically, O antigen is required for biofilm formation on the cholesterol-coated surface (gallstones) in serovar Typhimurium, Typhi, and Enteritidis. Surprisingly, curli-deficient mutants have been reported to have a considerable decrease of 45% in the biofilm compared with the wild type (Adcox et al., 2016). Extracellular DNA is also a contributor component in *Salmonella* biofilms matrix, derived from bacteria cell lysis within the biofilm. Typically, interacts with proteins or polysaccharides to stabilize the matrix; in *P. aeruginosa* and *S.* Typhimurium eDNA interacts with Pel polysaccharide and amyloid fibers, respectively (Adcox et al., 2016; Jennings et al., 2015).

Table 1. Main EPS of pathogenic bacteria biofilms

Bacteria	EPS produced	Reference
Staphylococcus aureus	Poly-N-acetylglucosamine surface polysaccharide Extracellular DNA	(Flannery et al., 2020; Kaplan et al., 2018)
Escherichia coli	Cellulose Poly-β-N-acetyl-D-glucosamine Curli protein Colanic acid	(Itoh et al., 2008; Limoli et al., 2015; Morgan et al., 2016; Smith et al., 2017)
Salmonella Typhimurium	Cellulose Curli protein Colanic acid	(Palomares-Navarro et al., 2022)
Pseudomonas aeruginosa	Alginate Psl Pel Lectins (LecA/LecB)	(Colvin et al., 2012; Passos da Silva et al., 2019)
Vibrio cholerae	Vibrio polysaccharide (VPS) Bap1 eDNA	(Berk et al., 2012; Colagiorgi et al., 2016; Schulze et al., 2021; Seper et al., 2011)
Listeria monocytogenes	Poly-β-N-acetylmannosamine Teichoic Acids BapL (biofilm associated protein)	(Colagiorgi et al., 2016)
Bacillus subtilis	Poly-N-acetylglucosamine	(Roux et al., 2015)
Acinetobacter baumannii	Poly-N-acetylglucosamine	(Flannery et al., 2020)
Clostridium perfringens	Type IV Pili	(Varga et al., 2008)

EPS composition is different depending on the bacteria; even the same one can produce different types of EPS depending on several factors such as the growth conditions: available nutrients, pH, salinity, temperature, etc. (Siddharth et al., 2021). For example, regarding nutrient composition, a study evaluated EPS composition in biofilms growing under a nutrient-enriched condition (nutrient ratio of 100:30:10 for carbon, nitrogen, and phosphorous, respectively), a phosphorus limitation (C: N: P ratio: 100:30:0), and a nitrogen limitation (C: N: P ratio: 100:0:10). The study revealed that biofilms grown under P-limited conditions were denser and with higher polysaccharide and eDNA content, whereas in nutrient-enriched and N-limited biofilms, the structure was more heterogeneous, with lower polysaccharide and eDNA (Desmond et al., 2018). On the other hand, Zhou et al. (2014) evaluated the effect of temperature; they reported higher concentrations of polysaccharides and protein content in *P. aeruginosa* biofilms grown at 20 °C instead of their optimal growth temperature (37 °C)

Virulence factors comprise a wide range of molecules produced by pathogenic microorganisms, enhancing their ability to evade their host defenses and cause disease; some of them include secreted products such as toxins, exopolysaccharides, enzymes, and cell surface structures such as capsules, lipopolysaccharides, glyco- and lipoproteins (Leitão, 2020). The EPS production in several bacterial species has been related to their pathogenicity; in some cases, it is considered a virulence factor. In this sense, several studies have been conducted to determine the influence of EPS on biofilm development and their pathogenicity; most of them have included the depletion of EPS-related genes.

A study evaluated the capacity of *S. mutans* to form biofilms after the depletion of genes involved in glucosyltransferase synthesis. Glucosyltransferases are responsible for glucan synthesis in *S. mutans* biofilms (causative of dental caries), and an affectation of this enzymatic complex can directly impact biofilm-forming ability. This was corroborated by Koo et al. (2010), who reported that *S. mutans* lacking genes *gtfB* (associated with insoluble glucan synthesis) and *gtfC* (associated with insoluble and soluble glucan synthesis) significantly reduced the ability of *S. mutants* to synthesize the EPS matrix and microcolony formation on saliva-coated hydroxyapatite surfaces. On the other hand, Pandit et al. (2020) evaluated the effect of EPS on *Bacillus subtilis* biofilm structure and viscoelasticity. Results demonstrated that *eps* mutants (depleted of the *eps* operon) showed reduced viscoelasticity, resulting in a very thin, fragile, and smooth biofilm.

On the other hand, Carter et al. (2016) evaluated the effect of depletion of curli fimbriae gene of lettuce- and clinical-isolated *E. coli* O157:H7 and their capacity to colonize baby spinach leaves compared with the wild-type strain. The bacterial count of curli-deletion mutant (lettuce-isolated) was reduced more than five-fold compared with the wild-type strain. A similar effect was observed in the clinical-isolated strain, resulting in a four-fold reduction of the counts in the mutant compared with the wild-type strain, evidencing the essential role of curli fimbriae for the *E. coli* O157:H7 attachment.

The biosynthetic machinery of polysaccharide intercellular adhesin (PIA) in *S. aureus* and *S. epidermidis* is encoded in the *ica* locus. Some studies have evidenced the role of PIA on bacterial pathogenicity. For example, it was reported that *ica*-positive *S. aureus* or *S. epidermidis* showed better *in-vivo* survival than their corresponding *ica* mutants in mouse infection models (Fluckiger et al., 2005). A study with *Pseudomonas* observed that the depletion of Pel and Psl exopolysaccharides had no impact on bacterial counts in wound-infected mice. Still, it reduced the capacity of *P. aeruginosa* to survive antibiotic treatment (200 μg/mL gentamicin sulfate) (Fleming et al., 2022).

It has become clear that EPS plays an important role in bacterial pathogenesis, which has been attributed to the role of this matrix in the development of biofilms and the acquired antibiotic resistance of human pathogens. Considering the ancestral use of plants as a treatment for bacterial infections, recent studies have focused on evaluating and exploiting the potential of phytochemicals to inhibit EPS synthesis.

2. Phytochemicals as Promising Compounds to Inhibit bacterial EPS

Biofilm EPS matrix supports bacterial adhesion to biotic and abiotic surfaces, contributes to aggregation, and functions as a three-dimensional scaffold that provides hydration, digestive capacity, and protection against antimicrobials, antibiotics, and host immune molecules (Flemming et al., 2016). As a result of the increased resistance acquired by bacteria as the cause of EPS synthesis and biofilm formation, antibacterial therapies became ineffective in eradicating biofilms, and therefore, developing new antibiofilm agents with new targets and mechanisms of action is necessary. Natural antibiofilm agents have emerged as a promising strategy for preventing biofilm development and other virulence factors synthesized by pathogenic bacteria. Plants are recognized as a source of bioactive compounds with high therapeutic potential, including antimicrobial activity, and recently, the exploration of their inhibitory activity in the biofilm formation of human pathogenic bacteria has remarkably increased (Adeosun et al., 2022). Particularly, phytochemicals (plant secondary metabolites) from medicinal plants, spices, fruits, and vegetables showed several advantages over synthetic molecules, including their green status and the different mechanisms of action, which could help to overcome the resistance problem associated with antibiotics (A. Borges et al., 2015). These compounds can affect biofilm formation at different levels, EPS inhibition being one of the most effective. The main phytochemical classes with reported high antibiofilm properties include phenolics, terpenoids, essential oils, coumarins, and alkaloids (Yong et al., 2019).

3. Molecular Docking to Study EPS-Related Proteins and Interactions with Phytochemicals

In recent years, molecular docking has been an applied method to identify potential molecules to inhibit the prevalence of microbial pathogens. This method can screen in a pool of phytochemicals existing in nature. Molecular docking studies can provide valuable insights into how phytochemical ligands interact with protein motifs of pathogens. These studies should be routinely conducted before screening for antibacterial activity, stability and safety (Abishad et al., 2021). Also, can be used to correlate *In vitro* results and explain the involved interactions (Awadelkareem et al., 2022). *In silico* analysis can be used to screen large libraries of molecules virtually, which can help to identify suitable antimicrobial candidates more quickly and cheaply than traditional methods.

Numerous research investigations established the antimicrobial capacity of plant extracts. However, more systematic studies of the antibiofilm potential are needed (Awadelkareem et al., 2022). Different groups of phytochemicals have been evaluated by their potential to inhibit EPS production in pathogenic bacteria. Among the inhibition mechanisms of EPS, is the interaction with the enzymes responsible for the synthesis of EPS. Other mechanisms related to EPS inhibition have been the interaction with receptor proteins of signaling molecules, avoiding cell-cell communications. Also, the interaction with adhesin proteins can reduce the capacity of bacteria to attach to surfaces (Abinaya & Gayathri, 2019; Ortega-Ramirez et al., 2020; Palomares-Navarro et al., 2022).

Quercetin has been evaluated to inhibit the LasR protein of *Pseudomonas aeruginosa*. LasR is associated with EPS production by inhibiting Pel and PsI exopolysaccharides. Also, it affects the interaction of signaling molecule N-octanoyl-DL-homoserine lactone (OHL) with LasR, avoiding cell-cell communication and further exopolysaccharide production. Quercetin at 40 μg/ml inhibited EPS production in *K. pneumoniae, P. aeruginosa*, and *Y. enterocolitica* by 80.39%, 73.52%, and 65.56%, respectively. Also, at 5–40 μg/mL reduced alginate production of *P. aeruginosa* in a range of 9–65%. Molecular docking analysis revealed that quercetin binds more rigidly with the LasR receptor protein (-7.20 Kcal/mol) than the signaling compound (-4.28 kcal/mol), acting as a potential competitive inhibitor. Additionally, the binding energy of quercetin increases in the presence of OHL in a non-competitive interaction. Interactions of quercetin with LasR were through one hydrogen bond and 15 hydrophobic interactions (Table 2) (Gopu, Meena, et al., 2015).

Packiavathy et al. (2012) evaluated the effect of a methanol extract of *Cuminum cyminum* at 2 mg/mL on EPS reduction in *Pseudomonas aeruginosa* PAO1 (70%), *P. mirabilis* (69%) *and S. marcescens* (63%). To estimate which molecule was the most active inhibitor, molecular docking of the chemical constituents of *Cuminum cyminum* extract with LasR protein, was performed. Methyl eugenol, lauric acid, linalool, and capric acid showed affinity energies of -27.54, -20.66, -17.20, and -17.18 kcal/mol, respectively, comparable with the natural ligand.

Adnan et al. (2020) reported an interaction between 28 phytochemicals identified in an extract of the medicinal plant *Adiantum philippense*. All compounds interacted with the active sites of different adhesin proteins IcsA (*S. flexneri)*, Sortase A (*S. aureus)*, OprD (*P. aeruginosa)*, EspA (*E. coli*), and FimH (uropathogenic *E. coli*) being rutin, scutellarin, and 18−β−glycyrrhetinic acid the compounds with the higher binding energies with the proteins (Table 2). Strong interactions with hydrogen binding pattern were formed with the residues Ala92 and Asn107 from Sortase A of *S. aureus*, Leu720 from IcsA of *S. flexneri,* Tyr121 from FimH of *E. coli*, Gln50, Glu66, and Glu67 from EspA of *E. coli*, and Phe133 and Arg39 from OprD of *P. aeruginosa*. Interactions of the phytochemicals with key residues could inhibit the crucial step of the biofilm-forming process, i.e., adhesion and impact with EPS production.

Cyanidin at the concentration of 150 μg/mL exhibited a 68.65% reduction in the exopolysaccharide production by *Klebsiella pneumoniae*. Molecular docking analysis revealed that cyanidin binds strongly with AHL receptor compared with the signaling molecule in competitive and non-competitive interactions. Quantification of precipitated EPS revealed 30.64% reduction in the EPS production in *K. pneumoniae* treated with 50 μg/mL of cyanidin compared to the control (Gopu & Shetty, 2016).

Table 2. Molecular docking analysis of phytochemical inhibitors of extracellular polymeric substances

Phytochemical	Bacteria	Protein	Dock score (kcal/mol)	Interactions		Reference
				H-bonds	Hydrophobic	
Cyanidin	*K. pneumoniae*	LasR	-7.76	Tyr47, Trp60, Thr75, Thr115	Leu40, Ala50, Leu125, Leu39, Ile52, Gly38, Gly126, Ala127, Leu36, Val76, Tyr56, Thr75, Tyr64, Asp73, Trp88, Ser129	(Gopu & Shetty, 2016)
Malvidin	*K. pneumoniae*	LasR	-8.84	-	-	(Gopu, Kothandapani, et al., 2015)
Petunidin	*K. pneumoniae*	LasR	-8.38	Tyr47	Ty64, Ile52, Leu39, Ala50, Gly38, Ala127, Val76, Leu36, Thr115, Thr75, Ser129, Leu110, Trp88, Tyr56, Asp73, Phe101	(Gopu et al., 2016)
Methyl eugenol	*P. aeruginosa*	LasR	-27.54	-	-	(Packiavathy et al., 2012)
Lauric acid			-20.66			
Linalool			-17.20			
Capric acid			-17.18			
Citral	*E. coli*	Glycosyltransferase	-5.8	-	-	(Ortega-Ramirez et al., 2020)
Geraniol	*E. coli*	Glycosyltransferase	-6.1	-	-	
Quercetin	*P. aeruginosa*	LasR	-7.2	Thr75	Tyr64, Tyr56, Asp73, Ser129, Leu36, Ala127, Val76, Ile52, Gly38, Ala50, Gly126, Cys79, Leu40, Leu39, Leu125	(Gopu, Meena, et al., 2015)
7-Trihydroxyflavone	*P. aeruginosa*	LasR	-8.03	Ser155	Val75, Leu85, Leu57, Tyr88, Tyr80, Ile99, Asp97	(Abinaya & Gayathri, 2019)

Phytochemical	Bacteria	Protein	Dock score (kcal/mol)	Interactions		Reference
				H-bonds	Hydrophobic	
Scutellarin	*S. aureus*	Sortase A	-7.3	Ala92, Asn107, Val168, Trp194	Ala104	(Adnan et al., 2020)
	E. coli	FimH	-8.2	Tyr21, Asp7, Val20		
	E. coli	EspA	-8.0	Lys57, Glu67, Arg183, Glu66	Lys54	
18–β–Glycyrrhetinic acid	*S. flexneri*	IcsA	-7.3	Asn703, Leu720, Gln722, Thr736	Lys721	
Rutin	*P. aeruginosa*	OprD	-9.5	Leu132, Phe133, Asp96, Asp41	Tyr26, Tyr97	
Linalool	*P. aeruginosa*	LasR	-6.1	-	Leu36, Leu40, Tyr47, Ala50, Ala70, Cys79, Leu125, Ala127, and Val76	(Alyousef et al., 2021)

7-Trihydroxyflavone (100 μg/mL) from *A. scholaris* leaf extract interacts with quorum sensing proteins (LasR, LasI and rhlR) and AHLase from *Pseudomonas aeruginosa* and inhibits EPS production 74.5%. Molecular docking revealed that TF had a strong affinity with LasR with a binding energy of -8.03 kcal/mol with hydrogen bonding with Ser155 (Abinaya & Gayathri, 2019). Gopu et al. (2016) evaluated the interaction of 48 phytochemicals with LasR receptor protein. Among these phytochemicals, petunidin had the highest -8.38 kcal/mol score. *In vitro* analysis showed a reduction of 64.41% in exopolysaccharide production of *K. pneumoniae*.

Ren et al. (2023) analyzed the interaction between the phytochemical isookanin with biofilm-formation-related proteins. Docking analysis revealed that isookanin forms hydrogen bonds with aminoacid residues of IcaR (Asp68, Gly75, Glu123), Aap A-domain (Lys359, Arg418, Asp596, Gly599), Aap b-domain (Asp87, Asn177, Glu186), Esp (Asp86, Ala87, Pro180, Thr182, Asn244), Bap b-domain (Gln612, Asp614, Lys616, Asn629, Asn650, Lys651), and Bap c (Asn120, Asp121, Thr123, Thr127, Thr130). The binding energies of the interactions were of -6.19, -7.63, -5.45, -7.58, -7.82, and -6.6 kcal/mol, respectively. The interaction of isookanin at 250 μg/mL decreased by 40.2% exopolysaccharide content. Additionally, it reduced 77.6% of eDNA production, EPS responsible for bacterial proliferation.

Gopu, Kothandapani, et al. (2015) evaluated the potential of a *Syzygium cumini* (L.) extract (SCE) to inhibit the EPS production in *Klebsiella pneumoniae*. SCE reduced 64.29% the EPS production and was attributed to malvidin compound present in the extract. Molecular docking showed an interaction of malvidin with LasR protein (related to quorum sensing) of -8.84 kcal/mol through hydrophobic interactions. Ortega-Ramirez et al. (2020) evaluated with molecular docking the interaction of terpenes citral and geraniol with glycosyltransferase (5EIY) to establish the mechanism associated with the inhibition of cellulose formation by *Escherichia coli* O157:H7. Docking analysis showed that the terpenes interacted with the hydrophobic pocket below the gatting loop and next to the finger helix of the glycosyltransferase enzyme with a binding energy of -5.8 and -6.1 kcal/mol, respectively. These regions are critical for translocating UDP-glucose to the nascent cellulose chain.

Anju et al. (2022) analyzed by molecular docking the interaction between hispidulin and baicalein with LasR to establish the mechanism associated with the inhibition of EPS. A reduction of 76.3 and 81.3% in exopolysaccharides was observed after treatment of 75 and 150 μg/mL of hispidulin and baicalein, respectively. Also, other virulence factors were reduced, such as alginate

(52.29 and 53.67%) and rhamnolipid production (79.84 and 84.6%) by hispidulin and baicalein, respectively, in *P. aeruginosa*. Evidence states that QS regulates the biofilm development and synthesis of EPS matrix components. The establishment of biofilm is mediated by the initial adherence to a target site and the development of biofilm structures by producing an extracellular matrix of specific biomolecules. Hispidulin and baicalein decreased the biofilm formation and subsequently affected the biosynthesis of matrix contents (alginate, exopolysaccharides, extracellular DNA, rhamnolipids.

Myrtus communis, a medicinal plant, decreased at sub-MIC concentrations the EPS production in *P. aeruginosa (*63%), *E. coli (*76%), *A. baumannii* (75%), and *S. marcescens* (66%). To establish the interactions associated with the inhibition, molecular docking of linalool, the bioactive compound of the extract, was performed. Linalool was docked with LasR protein and has an affinity in the same cavity of the natural ligand (3-oxo-C_{12}-acylhomoserine lactone) with a binding energy of -6.1 kcal/mol. The interactions linalool-LasR were hydrophobic and via Van der Waals forces (Alyousef et al., 2021).

An extract of the medicinal plant *Eruca sativa* Miller was evaluated for its capacity to reduce EPS production in different food-pathogen bacteria (Awadelkareem et al., 2022). *E. sativa* reduced the amount of EPS: 82.77% for *S. aureus*, 76.19% for *E. coli,* 70.42% for *P. aeruginosa*, and 59.73% for *S. enterica ser. Typhi* on the biofilm matrix. The interactions of the phytochemicals with higher affinity values (pyropheophorbide, sciadopitysin, rutin, and pheophorbide) were analyzed to correlate the obtained results. Docking analysis showed that phytochemicals of *E. sativa* interacted with the active sites of the adhesion proteins *Sortase A*, *EspA*, *OprD*, and *type IV b pilin* of *S. aureus*, *E. coli*, *P. aeruginosa*, and *S. enterica ser. Typhi*, respectively. Inhibition of adhesion proteins may help reduce the attachment capacity of the bacteria to surfaces and subsequently reduce the EPS formation.

4. Study of Molecular Dynamics of Phytochemicals and Bacterial Target Proteins Involved in Quorum Sensing Process Interactions

Molecular docking is a bioinformatics tool that predicts the best matching binding of a ligand to a protein, enhances the comprehension of structural orientation, and calculates the interaction energies in this binding state (Salmaso & Moro, 2018). However, the predictions of molecular docking often lack correlation with experimental results, primarily due to neglect of protein and ligand dynamics in its conformations, thermodynamics conditions, and solvation and ionic strength conditions in the molecular system (Pantsar & Poso, 2018). Consequently, bioinformatics software has emerged to validate the efficacy and interaction mechanisms proposed by molecular docking through molecular dynamics (MD) simulations software, among the most commonly employed are CHARMM (Brooks et al., 2009), AMBER (Case et al., 2005), NAMD (Phillips et al., 2005), and GROMACS (Van Der Spoel et al., 2005).

In MD simulations focused on ligand-protein systems, access is typically restricted to experimental protein structures deposited in the Protein Data Bank (PDB), while structure prediction using homology or comparative modeling techniques is also possible (Kuhlman & Bradley, 2019). Furthermore, repositories offer a significant number of target protein structures from pathogenic bacteria, and these structures can be employed in MD simulations to uncover novel knowledge into crucial structural features, pivotal for constraining their functionality. Therefore, the MD simulation techniques have substantiated their effectiveness in revealing insights to identify phytochemicals from plants, which could serve as antimicrobial inhibitors against enzymes implicated in the production of extracellular polymeric substances (EPS) and quorum sensing process involved in biofilm formation of pathogenic bacteria.

In this sense, it is important to mention that the analysis of MD trajectories in protein-ligand systems encompasses the extraction of key parameters, such as root mean square deviation (RMSD), root mean square fluctuation (RMSF), alterations in the hydrogen bonds quantities (Van Der Spoel et al., 2005), among others. The ligand-to-protein RMSD serves as a valuable parameter for evaluating the stability of a ligand-protein complex throughout MD simulations; it represents the variation in ligand position during the time of simulation expressed in distance (nm), relative to the initial docking position. The presence of hydrogen bonds can influence the stability of the ligand-protein complex, MD software can predict the quantity fluctuation of hydrogen bonds between protein and ligand and represents a significant tool for predicting the nature of the interactions that contribute to complex stability.

Furthermore, the RMSF analysis assesses the average fluctuation of each protein residue from MD trajectories, reveals the influence of ligand-protein interactions and highlights the residues implicated in ligand stability (Van Der

Spoel et al., 2005). However, a recent development in the field is MMPBSA, a bioinformatics software capable of predicting the energy binding within ligand-protein interactions (Valdés-Tresanco et al., 2021). Moreover, the decomposition analysis is another function of the MMPBSA to identify and quantify the impact (expressed in energy interactions) of each residue involved in ligand stability. The MMPBSA is a crucial bioinformatic tool for determining the most significant residues in ligand stability, aiding in identifying or modifying the ligands' pertinent structural characteristics. Besides, integrating bioinformatics tools, including MD simulation, has proven instrumental in advancing these efforts in the food and pharmacology sectors.

As a survival mechanism, biofilm formation is undoubtedly an important factor in the pathogenicity of certain bacteria, as it strategically enables their survival by adherence to solid surfaces, facilitating nutrient acquisition, and conferring resistance to antibiotics and antimicrobial agents. Beyond bacterial cells, the key constituents of biofilms are EPS, encompassing DNA, proteins, and polysaccharides of bacterial origin. Researchers have been focusing on disrupting the biofilm formation process utilizing plant phytochemical sources.

In this context, phytochemicals sourced from various plants, including phenolics, quinines, terpenoids, and organosulfur compounds, have been identified for their antimicrobial properties aimed at impeding biofilm formation. However, the antibiofilm effects of many phytochemical compounds primarily revolve around disrupting the quorum sensing process, which is crucial for coordinating colony-wide functions like biofilm development (Anabela Borges et al., 2016). Consequently, only a handful of plant-derived phytochemicals target the functionality of enzymes involved in EPS synthesis, and their effectiveness has been confirmed through MD simulation. Thus, the following section outlines recent studies dedicated to enhancing the inhibitory activity of plant phytochemicals on proteins involved in the quorum sensing process, validated through *in-vitro* and MD protocols.

A common application of MD simulation involves identifying the functional components within functional extracts, the functionality of which has been previously confirmed through experimental means. This approach was employed by (Singha et al., 2023) that identified the best-docked compounds (5,8-dimethoxycumarin, 3-methylsalicylhydrazide, cyclobarbital and orotic acid) with anti-biofilm activity; these were found in the methanolic extract of *Citrus limon*. Moreover, MD simulation revealed that these compounds exhibited an RMSD variation ranging from 0.05 to 0.3 nm, and regular hydrogen bonding patterns were also observed during simulation time (10 ns). These compounds may be responsible for the extract functionality, exhibiting the highest binding stability with proteins associated with biofilm formation in *E. coli* (FabH, FtsZ, RbmA, RmbC, SdiA and MreB).

Furthermore, this strategy has been employed in other studies as well, such as in the case of seed extract from *Psoralea corylifolia*, which demonstrated noteworthy quorum sensing inhibition in *Chromobacterium violaceum* (Husain et al., 2018). MD simulation revealed that bakuchiol, a compound in the extract, binds to the active pockets of LasR and RhlR. The analysis demonstrated the formation of 1–2 hydrogen bonds between bakuchiol and these regulatory proteins, suggesting a potential mechanism of interaction that interferes with the quorum sensing process. This observation provides valuable insights into the molecular basis of bakuchiol's inhibitory activity against LasR and RhlR, which are essential for virulence and biofilm formation in various pathogenic bacteria. Moreover, Shukla et al. (2021) also subjected LasR and RhlR from *P. aeruginosa* to MD simulation (100 ns) docked with 6-gingerol and curcumin, revealing that RMSD analysis of these complexes was identical to the control complex (LasR- *N*-(3-oxododecanoyl)-L-homoserine lactone).

Another study of MD simulation with LasR was performed by Gopu, Meena, et al. (2015); they evaluated the interaction dynamics between LasR and quercetin to assess the conformational changes and the interaction stability. *In vitro* assays showed the antibiofilm effect of quercetin to reduce alginate and exopolysaccharide production. Docking analysis demonstrated that quercetin binds more strongly to LasR than the signaling molecule (OHL). RMSD profile (20 ns run with the time step of 2 ps) showed that the protein-signaling molecule was unstable as compared with the protein-quercetin complex because it handles the activation of LasR protein, allowing the opening of the active cavity 307.6 $Å^3$. Conversely, quercetin closes the activation site by closing the loops in the cavity 179.8 $Å^3$.

MD simulations were employed to determine the antibiofilm mechanism of phytochemicals against S. aureus. For instance, Wu et al. (2019) demonstrate the remarkable stability of a natural phenolic compound, 3-*p*-trans-coumaroyl-2-hydroxyquinic acid, in its interaction with transpeptidase sortase A (SrtA), a protein involved in biofilm formation. The stability was evident from the RMSD analysis, which showed deviations around 0.15 nm. Furthermore, the MMPBSA analysis unveiled critical Gln113, Val168 and Arg197 for complex binding. These residues were found to be engaged in hydrogen bonding, contributing significantly to the binding energy with ΔE_{total} values of −7.9, −10.5, and 24.3 kJ mol^{-1}, respectively.

Moreover, Mao et al. (2023) simulated during 100 ns of LuxS from *S. aureus* with baicalein from *Scutellaria* roots, demonstrating that the RMSD remained below 0.5 and the RMSF analysis indicated minimal structural variations of the LuxS protein in the presence of baicalein compared to when without the ligand. This demonstrates the stability of the LuxS structure within the ligand-protein complex. The authors also conducted an MMPBSA analysis, revealing a free binding energy of −31.72±15.05 kcal/mol. However, further decomposition of the MMPBSA analysis by residue highlighted that the baicalein-LuxS complex is primarily influenced by Leu23 (−9.5336 kJ/moL) and Tyr95 (−7.5913 kJ/moL).

On the other hand, it is becoming increasingly evident that recent studies have showcased the antibiofilm activity of phytochemicals through MD simulation, primarily targeting proteins associated with the quorum sensing process while overlooking enzymes responsible for EPS formation. This discrepancy is reflected by the prevalence of literature reviews detailing plant extracts and phytochemicals exhibiting antibiofilm activity against a range of bacteria. However, a noticeable gap exists in the documentation of phytochemicals specifically displaying targeted inhibition of enzymes involved in the EPS formation (Anabela Borges et al., 2016; Ray & Pattnaik, 2023; Silva et al., 2023). Considering this, the utilization of MD simulation to investigate the inhibition of enzymes involved in EPS formation by phytochemicals represents an unexplored research area.

5. *In Vitro* Validation Studies of EPS Inhibition

Antimicrobials from natural sources are green alternatives to prevent and destroy biofilms, such as plant extracts and phytochemicals, not only in food processing but also in clinical settings (Sakarikou et al., 2020). Several classes of phytochemicals have shown to exhibit antimicrobial properties against clinically significant pathogens, such as *Enterobacter sp., S. aureus, K. pneumoniae, A. baumanii, P. aeruginosa* and *Enterococcus faecium* species, among others (Reza et al., 2019). Uncontrolled usage and excessive depletion of antibiotics have facilitated the rise of bacteria that are resistant to antibiotics, commonly known as antibiotic- strains or superbugs, causing severe infections worldwide (Mitra et al., 2022). These bacteria possess a range of harmful factors, a high level of antibiotic resistant resistance, and the ability to form biofilms, contributing to their classification as challenging-to-manage pathogens (Patil et al., 2021). Phytochemicals have been shown to be effective against the development of antimicrobial resistance and the prevention and control of biofilms. These compounds interfering with key factors involved in biofilm development, such as EPS production, motility, adhesion, intercellular aggregation, and communication (Sakarikou et al., 2020). Specifically, EPS have been inhibited by different classes of phytochemicals with diverse chemical structures (Figure 2), which gives them the ability to have different inhibition targets compared to antibiotics. Using phytochemicals against EPS synthesis during biofilm development has increased the sensibility of bacteria to antibiotics (Khare et al., 2021).

Phenolic compounds are secondary metabolites synthesized through the shikimic acid and phenylpropanoid pathways. They are chemically defined as compounds containing hydroxylated aromatic rings with a hydroxy group being attached directly to the phenyl, substituted phenyl, or another aryl group (de la Rosa et al., 2019). They are found in most plant tissues, including fruits and vegetables and possess numerous bioactive properties, including antimicrobial and antibiofilm. Major phenolic compounds include phenolics, phenolic acids, flavonoids, coumarins, stilbenes, tannins, lignans, and lignins (de la Rosa et al., 2019).

Vainillin and syringic acid at 1/20 MIC concentration reduced 60% EPS polysaccharide production, with the concomitant inhibition of different strains of *S. epidermidis* biofilm formation by 80% (Minich et al., 2022). Bernal-Mercado et al. (2020) evaluated the combined effect of catechin (0·00005 mol/L), vanillic (0·00074 mol/L) and protocatechuic acids (0·00162 mol/L) on uropathogenic *E. coli* (UPEC) biofilm matrix synthesis. The authors observed that the combination reduced 20–40% of the carbohydrate and protein content in the biofilm matrix, and curli fimbriae, a main protein component of UPEC biofilm matrix, were affected.

Table 2. Chemical structure of plant bioactives capable of inhibiting EPS production in pathogenic bacteria

Compound	Chemical Structure	Reference
Terpenes and terpenoids		
Carvacrol	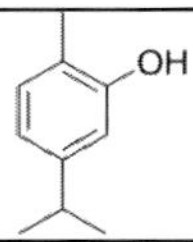	(Liu et al., 2021)
Eugenol	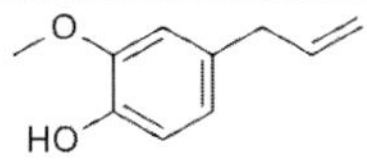	(B. Li et al., 2022)
α-terpinene	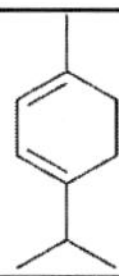	(Adeosun et al., 2022)
Phytol	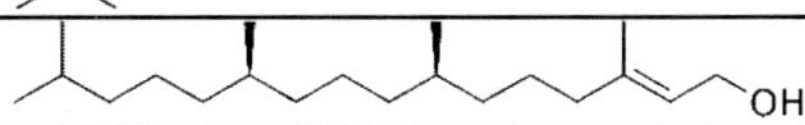	(Srinivasan et al., 2017)
Citral	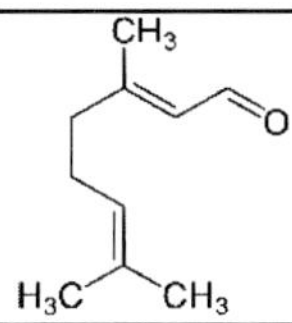	(Ortega-Ramirez et al., 2020)
Geraniol	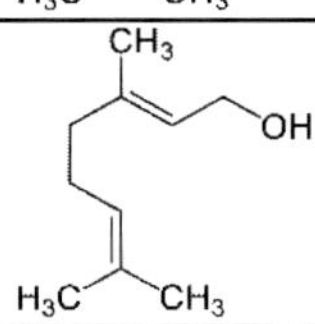	(Ortega-Ramirez et al., 2020)
Phenolic acids		
Protocatechuic acid	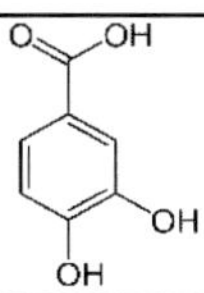	(Bernal-Mercado et al., 2020)
Vainillin	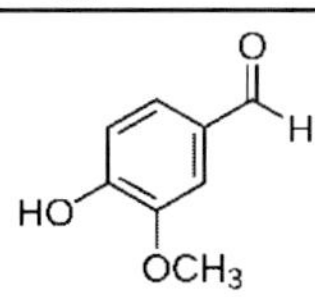	(Minich et al., 2022)
Gallic acid	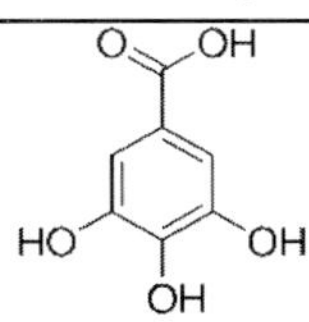	(Kang et al., 2018))

Table 2. (Continued)

Compound	Chemical Structure	Reference
Flavonoids		
Quercetin	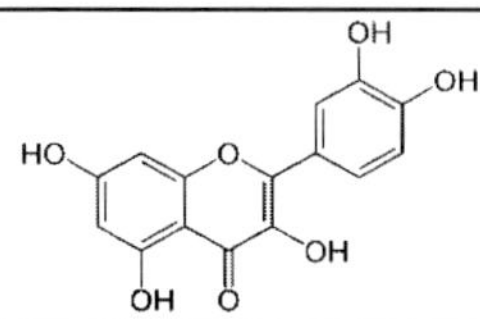	(Vazquez-Armenta et al., 2018)
Morin	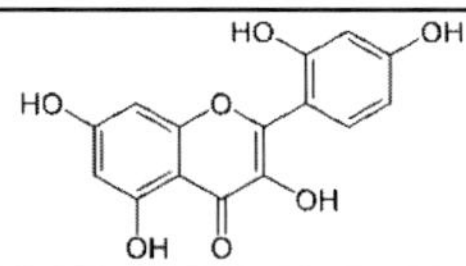	(Chemmugil et al., 2019)
Epigallocatechin gallate	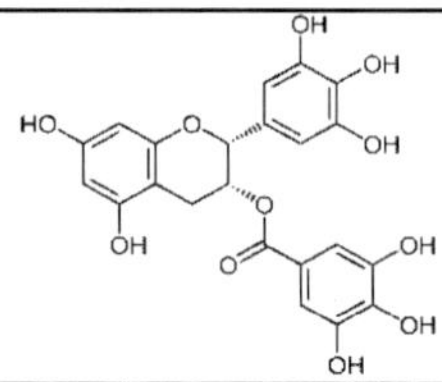	(Schneider-Rayman et al., 2021)

Staphylococcus aureus is one of the most frequent causes of biofilm-associated infections, especially methicillin-resistant *S. aureus* (MRSA). Mastoor et al. (2022) investigated the antimicrobial and anti-biofilm properties of natural compounds (salicylaldehyde, N,N-dimethyl-cinnamaldehyde, α-methyl-cinnamic acid, vanillin, α-methyl-trans-cinnamaldehyde, α-bromo-trans-cinnamaldehyde, 4-acetoxy-3-methoxycinnamaldehyde, 3-hydroxy-4-methoxycinnamic acid, 4-hydroxy-3,5-dimethoxycinnamic acid, and trans-4-nitrocinnamic acid) against methicillin-resistant *Staphylococcus aureus* (MRSA) isolates. The MIC's of the natural compounds ranged from 0.75 to 160 mg/mL. Salicylaldehyde, α-methyl-trans-cinnamaldehyde, trans-4-nitro cinnamic acid, and vanillin showed significant biofilm inhibition/reduction activity (15-92%) at 1-10 mg/mL. Gene expression analysis revealed that salicylaldehyde, α-methyl-trans-cinnamaldehyde, and α-bromotrans-cinnamaldehyde significantly down-regulated the expression of biofilm-related genes *icaA*, *clfA*, and *fnbA* compared to the untreated resistant isolate. In another study, gallic acid inhibited by 13.27-86.83% the *E. coli* biofilm formation at concentrations ranging between 0.25-4.00 mg/mL. EPS components (polysaccharide, protein and DNA) were significantly reduced in treated *E. coli* (0.25 mg/mL gallic acid) compared to control bacteria. The expression of *pgaABCD* genes that encoded PGA, the main polysaccharide component in *E. coli* biofilm matrix, were suppressed under the 2 mg/mL gallic acid, evidencing the potential of this compound to inhibit EPS synthesis and the concomitant biofilm formation (Kang et al., 2018).

In the case of flavonoids, a recent study evaluated the preventive effect of quercetin against *L. monocytogenes* biofilms. This compound, at a subinhibitory concentration, reduced the total extracellular protein content by 41%, affecting *Listeria* surface colonization and the development of mature biofilms. Exopolysaccharides and eDNA were also quantified, but quercetin did not affect these EPS components (Vazquez-Armenta et al., 2018). In a study carried out by Chemmugil et al. (2019), it was found that morin showed low antibacterial activity against resistant strains of *S. aureus* (50-500 μg/mL). However, at 100 μg/mL it exhibited antibiofilm activity against clinically derived MRSA and VRSA; and a significant biofilm inhibition (55-65%), reduced motility and spread (76-86%). Morin also reduced the EPS production from both resistant strains; control of MRSA, VRSA produced 8.2 and 3 mg/mL, respectively, and the production of EPS was reduced to 6.28 and 0.6 mg/mL for MRSA and VRSA, respectively. Pathogenicity and virulence of microorganisms are related to EPS production and are crucial for bacterial adhesion, adaptation, and protection. The inhibition of morin was understood by attenuation of gliding motility, propagation, and EPS production, which is essential for propagation and biofilm formation. Morin can also interfere with the interaction of *Staphylococcus* biofilm on the glass surface.

Coumarins are a large family of heterocyclic compounds whose chemical structure comprises fused benzene and pyrone rings. These are secondary metabolites commonly found in high plants and used as fragrances. Coumarins have also been tested as potential inhibitors of biofilm formation and other virulence factors such as EPS. Thakur et al. (2020) determined the antimicrobial, antiquorum and antibiofilm activity of coumarin (2 mM) against *S.* Typhimurium. Results showed that coumarin reduced biofilm formation and attenuated curli and cellulose production, likely by down-regulating the transcript levels of major biofilm formation genes *csgD*, *csgA* and *adrA*.

The antimicrobial efficacy of many herbs and spices is primarily due to their abundant content of essential oils. Essential oils are natural, volatile, and aromatic liquids extracted from aromatic and medicinal plants; they are a complex mixture of secondary metabolites (terpenes, phenolic compounds, alcohol) and possess a wide range of biological activities, including antioxidant, antiinflammatory, antimutagenic, and antimicrobial (Falleh et al., 2020).

Clove essential oil influenced biofilm formation by *L. monocytogenes* and effectively removed biofilms from plant surfaces (C. Zhang et al., 2020). The findings indicated that all concentrations tested (0.5, 1.0, and 2.0 mg/mL) reduced 35% to 88% in extracellular polysaccharides and 34% to 76% in protein content within *Listeria* biofilms. Furthermore, this essential oil played a role in reducing the expression of *agrA*, *agrD*, *agrC*, and *prfA* genes, involved in quorum sensing and biofilm formation in *L. monocytogenes*. Clove essential oil at a concentration of 1.2 mg/mL reduced *S.* Typhimurium adhering to stainless steel surfaces to 1.8 CFU/cm² (Silva-Espinoza et al., 2020). Moreover, the essential oils from *Origanum vulgare* (at 0.25%) and *Rosmarinus officinalis* (at 4%) effectively eradicated both nascent and developed biofilms formed by *S.* Enteritidis on stainless steel surfaces, showcasing a time-dependent impact (Lira et al., 2020).

Hakimi Alni et al. (2020) found that *Cuminum cyminum*-derived essential oil, composed mainly of terpenes such as terpineol, carene, and pinene, prevented the growth of *S.* Typhimurium at a concentration of 2.62 μL/mL. At concentrations below those that completely inhibit growth, this essential oil also decreased biofilm formation and caused cells to detach from the biofilms. In contrast, control bacteria adhered tightly within a dense matrix. In addition, these researchers documented that the subinhibitory level of *C. cyminum* oil down-regulated important genes related to EPS formation, such as genes related to cellulose synthesis (*csgD* and *adrA*), as well as quorum sensing (*sdiA* and *luxS*). These results are important because curli fimbriae and cellulose are the main extracellular polymeric substances in *Salmonella* biofilms, both produced in response to environmental stress (Mastoor et al., 2022).

The antibacterial and antibiofilm activity of essential oils is associated with the combination of chemical compounds; specifically, terpenes and terpenoids present in essential oils have shown significant efficacy against a wide range of microorganisms, even at concentrations below minimal inhibitory levels. Terpenes belong to the biggest class of secondary metabolites and consist of five carbon isoprene units assembled together. Terpenes are simple hydrocarbons, while terpenoids contain different functional groups and oxidized methyl groups (Wilbon et al., 2013).

Ortega-Ramirez et al. (2020) evaluated the effect of citral and geraniol, the main compounds of lemongrass essential oil, on glucan production of *E. coli* O157:H7 and results showed a significant reduction at concentrations of 0.5 and 0.25 mg/mL at 12 h of incubation at 37 °C. Additionally, enzymatic assays showed that these compounds inhibited the activity (uncompetitive inhibition) of the glucosyltransferase enzyme, responsible for glucan synthesis, at IC_{50} of 8.5 and 6.5 µM, respectively.

Similarly, M. Li et al. (2012) investigated the impact of eugenol on glucan production and biofilm formation in *Streptococcus sobrinus*, a pathogen associated with dental caries. The findings revealed that the minimum biofilm-inhibiting concentration of eugenol in an artificial mouth model was 0.625 mg/mL, while the minimum concentration of the artificial mouth model to inhibit biofilm was 8 mg/mL. There was a 63% decrease in the synthesis of insoluble glucans and a 46% decrease in soluble glucans. Glucans play a fundamental role in the formation of biofilms of caries-causing bacteria. However, the precise mechanisms by which this terpene suppresses EPS synthesis are not explicitly described. Eugenol may hinder enzymes involved in glucan synthesis, such as glucosyltransferase. Furthermore, several studies employing this approach focus on oral bacteria.

Carvacrol effectively inhibited biofilm formation and EPS production from *E. cloacae*. The MIC of carvacrol oil against *E. cloacae* was 256 µg/mL. Bacterial cell motility was inhibited at subinhibitory concentrations of 64 and 128 µg/mL, and biofilm thickness and extracellular polymeric matrix excretion were reduced, demonstrated by scanning electron microscopy and confocal laser scanning microscopy. Additionally, the correlation indicated that the inhibition of biofilms in mature stages was related to the decrease in EPS production (Liu et al., 2021).

The phytochemicals α-terpinene, phytol, camphene, fisetin, and glycitein were evaluated for their capacity to inhibit the synthesis of exopolysaccharides, curli fibers, and hyper-muco-viscosity of the carbapenem-resistant and extended-spectrum beta-lactamase-positive *K. pneumoniae* strains. Results showed an exopolysaccharide reduction of up to 65.91% in phytol and camphene treatments, whereas curli expression was inhibited by phytol, glycitein, and fistein at 0.5 and 1.0 mg/mL. Additionally, 1 mg/mL of phytol, glycitein and fistein significantly reduced *K. pneumonie* hyper-muco-viscosity (Adeosun et al., 2022). Treatment with phytol (5 and 10 µg/mL) against *Serratia marcescens* showed decreasing levels of biofilm formation, confirmed by microscopic analysis. This was related to the inhibition of swarm motility and the production of EPS. This compound also down-regulated the *fimA*, *fimC*, *flhC*, *flhD*, *bsmB*, *pigP*, and *shlA* gene expressions (associated to fimbriae synthesis and motility in *S. marcescens*), and reduced the level of virulence enzymes such as lipase and protease productions (Srinivasan et al., 2017). In another investigation, this compound was tested at three sub-MIC concentrations: 0.5, 0.25, and 0.125 MIC, against

P. aeruginosa causing a reduction in biofilm formation in the range of 74.00 to 84.33%. Phytol (0.5 MIC) also effectively reduced *P. aeruginosa* twitching and flagella motility, this effect is related to the reduction in EPS synthesis. In fact, the treated bacteria could not produce a twitching zone and had almost round, smooth and regular colony edges (Pejin et al., 2015).

In another study, D-limonene was used to inhibit biofilms of *E. coli* O157:H7 where it was observed that this compound at 1.25% and 2.5% v/v, inhibited the formation of *E. coli* biofilms by 55%, and 92%, respectively. This effect was reflected in the suppression of curli and a significant decrease in the production of EPS in cells exposed to D-limonene by 35% and 60% at concentrations of 1. 25 and 2.5% (v/v), respectively (Wang et al., 2018). Allicin (10 and 128 μg/mL), an organosulfur compound found in garlic, reduced *P. aeruginosa* early bacterial adhesion, reduced EPS secretion in a dose-dependent manner, and down-regulate virulence factors' production such as exotoxin A and elastase, and completely inhibited production of rhamnolipid and pyoverdin.

6. Applied Strategies for EPS Disruption

The recalcitrant and harmful nature of biofilms led to the use of phytochemicals to inhibit EPS synthesis, affect the biofilm formation process, or destabilize established biofilms through disruption of the EPS matrix. Besides the well-known antibiofilm properties of phytochemicals demonstrated *in silico* and *in vitro* studies, using these compounds alone or in carrier materials applied in food systems or clinical settings became necessary.

Clinical or food industry surfaces are hotspots susceptible to contamination by biofilm formation by ambient conditions such as temperature, nutrient source, relative humidity, etc. Few strategies have effectively eradicated biofilm formation which conduce to the health risk of pathogen bacteria. Among the techniques that have been used are the application of high concentrations of sanitizers and antibiotics carrying bacterial resistance.

Phytochemicals have proven effective in reducing EPS synthesis and biofilm formation. Kim et al. (2015) reported that topical application of cranberry flavonoids (twice-daily, 8 a.m. and 6 p.m. by 5 min exposure) on hydroxyapatite surfaces significantly reduced the insoluble EPS content (>80%) and hindered *S. mutans* growth within mixed-species biofilms. This effect was corroborated by the 3D architecture of cranberry-treated biofilms, which was severely compromised, showing a defective EPS-matrix and inhibition of microcolony formation on the saliva-coated hydroxyapatite surfaces. Additionally, topical applications of cranberry flavonoids weaken the mechanical stability of the biofilms, allowing the removal of almost 90% of the biofilm from the hydroxyapatite surface when applying a shear stress of 0.449 N/m^2.

The antibiofilm effect of a grapefruit seed extract (GSE) against biofilm-forming strains of *Staphylococcus aureus* and *Escherichia coli* has been evaluated on a stainless-steel surface (Song et al., 2019). Biofilms were degraded after GSE exposure and could be associated with in vitro analysis that showed that GSE reduced EPS production (28.5%) and bacteria motility. GSE significantly degrades biofilms on stainless steel at concentrations above the MIC's (25 µg/mL: *S. aureus;* 250 µg/mL: *E. coli*). The SEM images of biofilm cells attached to stainless steel coupons demonstrate that when GSE concentration increases, the attachment and viability of cells decreases.

Y. Zhang et al. (2022) evaluated the effect of chlorophyllin and chlorophyllin-based photodynamic inhibition on *Pseudomonas reinekei* and *Pseudomonas palleroniana* biofilms in fresh-cut Pakchoi during quality storage. The treatments reduced the biofilm formation by inhibiting extracellular polysaccharides and extracellular proteins *in vitro and in vivo.* The Chl-based PDI increased the visual quality and the content of total soluble solids, SOD activity, chlorophyll content and reduced yellowness in the cruciferous.

B. Li et al. (2022) evaluated the effect of the stilbenoid rhapontigenin against the virulence factors of *Pectobacterium carotovorum* inoculated (at 0 MIC, 1/8 MIC, 1/4 MIC and 1/2 MIC) in Chinese cabbage. Rhapontigenin at ½ MIC has the best inhibition of the virulence of *P. carotovorum*, reducing the Chinese cabbage's decay rate. The reduction was associated with the potential as a quorum-sensing inhibitor. As QS, the treatment with rhapontigenin mediates EPS synthesis and may inhibit the QS system, resulting in reduced synthesis of EPS or preventing their production.

Nanoencapsulation may be a promising alternative to carrying active compounds into the biofilm matrix for further release. Polymeric nanoparticles assembled with tannic acid and biguanide-based polymetformin at 8–32 µg/mL were used to eradicate MRSA biofilms, and it was observed inhibition of biofilm formation *in vitro* (~2 $\log_{10}$). Additionally, tannic acid-nanoparticles suppressed infection in the excisional wound model in murine models by reducing biofilm-associated MRSA by 1.8 $\log_{10}$ reduction of biofilm-associated MRSA better than vancomycin. These demonstrated the capacity of this compound to disrupt MRSA EPS matrix and biofilm (Li et al., 2020).

Conclusion

Biofilm human infections are one of the major concerns in world health due to the increased antibiotic resistance acquired in this bacterial mode of growth. This is mainly attributed to EPS of the biofilm matrix, which creates a physicochemical barrier that prevents the entry of antibiotics to the bacteria embedded in the biofilm. There are extensive research evidence that supports the biofilm inhibitory effect of phytochemicals, as well as their impact on EPS synthesis. The comprehensive study of the inhibitory effect of phytochemicals on EPS and biofilm formation involving in vitro, in silico and application studies is necessary to understand the antibiofilm effect of phytochemicals and broaden the search for potential antivirulence compounds. Among these, in silico studies have shown to be a key tool to elucidate the molecular interactions and dynamics between phytochemicals and key proteins in the synthesis of EPS, opening the panorama to the possible inhibitory mechanisms of action of phytochemicals.

References

Abinaya, M., & Gayathri, M. (2019). Inhibition of biofilm formation, quorum sensing activity and molecular docking study of isolated 3, 5, 7-Trihydroxyflavone from *Alstonia scholaris* leaf against *P. aeruginosa*. *J Bioorganic Chemistry, 87*, 291-301.

Abishad, P., Niveditha, P., Unni, V., Vergis, J., Kurkure, N. V., Chaudhari, S., Rawool, D. B., & Barbuddhe, S. B. (2021). *In silico* molecular docking and in vitro antimicrobial efficacy of phytochemicals against multi-drug-resistant enteroaggregative *Escherichia coli* and non-typhoidal *Salmonella* spp. *J Gut pathogens, 13*(1), 1-11.

Adcox, H. E., Vasicek, E. M., Dwivedi, V., Hoang, K. V., Turner, J., & Gunn, J. S. (2016). Salmonella extracellular matrix components influence biofilm formation and gallbladder colonization. *Infection immunity, 84*(11), 3243-3251.

Adeosun, I. J., Baloyi, I. T., & Cosa, S. (2022). Anti-biofilm and associated anti-virulence activities of selected phytochemical compounds against *Klebsiella pneumoniae*. *Plants, 11*(11), 1429.

Adnan, M., Patel, M., Deshpande, S., Alreshidi, M., Siddiqui, A. J., Reddy, M. N., Emira, N., & De Feo, V. (2020). Effect of *Adiantum philippense* extract on biofilm formation, adhesion with its antibacterial activities against foodborne pathogens, and characterization of bioactive metabolites: An *in vitro-in silico* approach. *Frontiers in microbiology, 11*, 823.

Alyousef, A. A., Husain, F. M., Arshad, M., Ahamad, S. R., Khan, M. S., Qais, F. A., Khan, A., Alqasim, A., Almutairi, N., & Ahmad, I. (2021). *Myrtus communis* and its bioactive phytoconstituent, linalool, interferes with Quorum sensing regulated virulence functions and biofilm of uropathogenic bacteria: *In vitro and in silico* insights. *Journal of King Saud University-Science, 33*(7), 101588.

Anju, V., Busi, S., Mohan, M. S., Ranganathan, S., Ampasala, D. R., Kumavath, R., & Dyavaiah, M. (2022). *In vivo, in vitro* and molecular docking studies reveal the anti-virulence property of hispidulin against *Pseudomonas aeruginosa* through the modulation of quorum sensing. *International Biodeterioration & Biodegradation, 174*, 105487.

Awadelkareem, A. M., Al-Shammari, E., Elkhalifa, A. O., Adnan, M., Siddiqui, A. J., Mahmood, D., Azad, Z. A. A., Patel, M., Mehmood, K., & Danciu, C. (2022). Anti-adhesion and antibiofilm activity of *Eruca sativa* miller extract targeting cell adhesion proteins of food-borne bacteria as a potential mechanism: Combined *in vitro-in silico* approach. *Plants, 11*(5), 610.

Berk, V., Fong, J. C., Dempsey, G. T., Develioglu, O. N., Zhuang, X., Liphardt, J., Yildiz, F. H., & Chu, S. (2012). Molecular architecture and assembly principles of *Vibrio cholerae* biofilms. *Biophysical Journal, 102*(3), 617a.

Bernal-Mercado, A. T., Gutierrez-Pacheco, M. M., Encinas-Basurto, D., Mata-Haro, V., Lopez-Zavala, A. A., Islas-Osuna, M. A., Gonzalez-Aguilar, G. A., & Ayala-Zavala, J. F. (2020). Synergistic mode of action of catechin, vanillic and protocatechuic acids to inhibit the adhesion of uropathogenic *Escherichia coli* on silicone surfaces. *Journal of Applied Microbiology, 128*(2), 387-400.

Borges, A., Abreu, A. C., Dias, C., Saavedra, M. J., Borges, F., & Simões, M. (2016). New perspectives on the use of phytochemicals as an emergent strategy to control bacterial infections including biofilms. *Molecules, 21*(7), 877.

Borges, A., J Saavedra, M., & Simoes, M. J. (2015). *Insights on antimicrobial resistance, biofilms and the use of phytochemicals as new antimicrobial agents.* 22(21), 2590-2614.

Carter, M. Q., Louie, J. W., Feng, D., Zhong, W., & Brandl, M. (2016). Curli fimbriae are conditionally required in *Escherichia coli* O157: H7 for initial attachment and biofilm formation. *Food microbiology, 57*, 81-89.

Case, D. A., Cheatham III, T. E., Darden, T., Gohlke, H., Luo, R., Merz Jr, K. M., Onufriev, A., Simmerling, C., Wang, B., & Woods, R. J. (2005). The Amber biomolecular simulation programs. *Journal of computational chemistry, 26*(16), 1668-1688.

Chemmugil, P., Lakshmi, P., & Annamalai, A. (2019). Exploring Morin as an anti-quorum sensing agent (anti-QSA) against resistant strains of *Staphylococcus aureus*. *Microbial pathogenesis, 127*, 304-315.

Colagiorgi, A., Di Ciccio, P., Zanardi, E., Ghidini, S., & Ianieri, A. (2016). A look inside the *Listeria monocytogenes* biofilms extracellular matrix. *Microorganisms, 4*(3), 22.

Colvin, K. M., Irie, Y., Tart, C. S., Urbano, R., Whitney, J. C., Ryder, C., Howell, P. L., Wozniak, D. J., & Parsek, M. R. (2012). The Pel and Psl polysaccharides provide

Pseudomonas aeruginosa structural redundancy within the biofilm matrix. *Environmental microbiology, 14*(8), 1913-1928.

de la Rosa, L. A., Moreno-Escamilla, J. O., Rodrigo-García, J., & Alvarez-Parrilla, E. (2019). Phenolic compounds. In *Postharvest physiology and biochemistry of fruits and vegetables* (pp. 253-271): Elsevier.

Desmond, P., Best, J. P., Morgenroth, E., & Derlon, N. (2018). Linking composition of extracellular polymeric substances (EPS) to the physical structure and hydraulic resistance of membrane biofilms. *Water research, 132*, 211-221.

Di Martino, P. (2018). Extracellular polymeric substances, a key element in understanding biofilm phenotype. *AIMS microbiology, 4*(2), 274.

Falleh, H., Jemaa, M. B., Saada, M., & Ksouri, R. (2020). Essential oils: A promising eco-friendly food preservative. *Food Chemistry, 330*, 127268.

Flannery, A., Le Berre, M., Pier, G. B., O'Gara, J. P., & Kilcoyne, M. (2020). Glycomics microarrays reveal differential in situ presentation of the biofilm polysaccharide poly-N-acetylglucosamine on *Acinetobacter baumannii* and *Staphylococcus aureus* cell surfaces. *International journal of molecular sciences, 21*(7), 2465.

Fleming, D., Niese, B., Redman, W., Vanderpool, E., Gordon, V., & Rumbaugh, K. P. (2022). Contribution of *Pseudomonas aeruginosa* exopolysaccharides Pel and Psl to wound infections. *Frontiers in Cellular Infection Microbiology, 12*, 835754.

Flemming, H.-C., Neu, T. R., & Wingender, J. (2016). *The perfect slime: microbial extracellular polymeric substances (EPS)*: IWA publishing.

Fluckiger, U., Ulrich, M., Steinhuber, A., Döring, G., Mack, D., Landmann, R., Goerke, C., & Wolz, C. (2005). Biofilm formation, icaADBC transcription, and polysaccharide intercellular adhesin synthesis by staphylococci in a device-related infection model. *Infection & immunity, 73*(3), 1811-1819.

Gopu, V., Kothandapani, S., & Shetty, P. H. (2015). Quorum quenching activity of *Syzygium cumini* (L.) Skeels and its anthocyanin malvidin against *Klebsiella pneumoniae*. *Journal of food science, 79*, 61-69.

Gopu, V., Meena, C. K., Murali, A., & Shetty, P. H. (2016). Petunidin as a competitive inhibitor of acylated homoserine lactones in *Klebsiella pneumoniae*. *RSC advances, 6*(4), 2592-2601.

Gopu, V., Meena, C. K., & Shetty, P. H. (2015). Quercetin influences quorum sensing in food borne bacteria: in-vitro and in-silico evidence. *PLoS one, 10*(8), e0134684.

Gopu, V., & Shetty, P. H. (2016). Cyanidin inhibits quorum signalling pathway of a food borne opportunistic pathogen. *Journal of food science & technology, 53*, 968-976.

Gunn, J. S., Bakaletz, L. O., & Wozniak, D. J. (2016). What's on the outside matters: the role of the extracellular polymeric substance of gram-negative biofilms in evading host immunity and as a target for therapeutic intervention. *Journal of Biological Chemistry, 291*(24), 12538-12546.

Hakimi Alni, R., Ghorban, K., & Dadmanesh, M. (2020). Combined effects of Allium sativum and *Cuminum cyminum* essential oils on planktonic and biofilm forms of *Salmonella typhimurium* isolates. *Biotech, 10*, 1-10.

Hu, Y., Ruan, X., Lv, X., Xu, Y., Wang, W., Cai, Y., Ding, M., Dong, H., Shao, J., & Yang, D. (2022). Biofilm microenvironment-responsive nanoparticles for the treatment of bacterial infection. *Nano Today, 46*, 101602.

Husain, F. M., Ahmad, I., Khan, F. I., Al-Shabib, N. A., Baig, M. H., Hussain, A., Rehman, M. T., Alajmi, M. F., & Lobb, K. A. (2018). Seed extract of *Psoralea corylifolia* and its constituent bakuchiol impairs AHL-based quorum sensing and biofilm formation in food-and human-related pathogens. *Frontiers in cellular and infection microbiology, 8*, 351.

Itoh, Y., Rice, J. D., Goller, C., Pannuri, A., Taylor, J., Meisner, J., Beveridge, T. J., Preston III, J. F., & Romeo, T. (2008). Roles of pgaABCD genes in synthesis, modification, and export of the *Escherichia coli* biofilm adhesin poly-β-1, 6-N-acetyl-D-glucosamine. *Journal of bacteriology, 190*(10), 3670-3680.

Jennings, L. K., Storek, K. M., Ledvina, H. E., Coulon, C., Marmont, L. S., Sadovskaya, I., Secor, P. R., Tseng, B. S., Scian, M., & Filloux, A. (2015). Pel is a cationic exopolysaccharide that cross-links extracellular DNA in the *Pseudomonas aeruginosa* biofilm matrix. *Proceedings of the National Academy of Sciences, 112*(36), 11353-11358.

Kang, J., Li, Q., Liu, L., Jin, W., Wang, J., & Sun, Y. (2018). The specific effect of gallic acid on *Escherichia coli* biofilm formation by regulating pgaABCD genes expression. *Applied microbiology and biotechnology, 102*, 1837-1846.

Kaplan, J. B., Mlynek, K. D., Hettiarachchi, H., Alamneh, Y. A., Biggemann, L., Zurawski, D. V., Black, C. C., Bane, C. E., Kim, R. K., & Granick, M. S. (2018). Extracellular polymeric substance (EPS)-degrading enzymes reduce staphylococcal surface attachment and biocide resistance on pig skin in vivo. *PLoS one, 13*(10), e0205526.

Khare, T., Anand, U., Dey, A., Assaraf, Y. G., Chen, Z.-S., Liu, Z., & Kumar, V. (2021). Exploring phytochemicals for combating antibiotic resistance in microbial pathogens. *Frontiers in pharmacology, 12*, 720726.

Kim, D., Hwang, G., Liu, Y., Wang, Y., Singh, A. P., Vorsa, N., & Koo, H. (2015). Cranberry flavonoids modulate cariogenic properties of mixed-species biofilm through exopolysaccharides-matrix disruption. *PLoS one, 10*(12), e0145844.

Koo, H., Xiao, J., Klein, M., & Jeon, J. (2010). Exopolysaccharides produced by *Streptococcus mutans* glucosyltransferases modulate the establishment of microcolonies within multispecies biofilms. In: Am Soc Microbiol.

Kuhlman, B., & Bradley, P. (2019). Advances in protein structure prediction and design. *Nature Reviews Molecular Cell Biology, 20*(11), 681-697.

Leitão, J. H. (2020). Microbial virulence factors. *International journal of molecular sciences, 21*(15), 5320.

Li, B., Huang, J., Yi, Y., Liu, S., Liu, R., Xiao, Z., & Li, C. (2022). Effects of rhapontigenin as a novel quorum-sensing inhibitor on exoenzymes and biofilm formation of *Pectobacterium carotovorum* subsp. carotovorum and its application in vegetables. *Molecules, 27*(24), 8878.

Li, M., Lai, G., Wang, J., & Ye, D. (2012). The inhibition of eugenol on glucan is essential for the biofilm eradication effect on caries-related biofilm in an artificial mouth model. *Natural product research, 26*(12), 1152-1155.

Limoli, D. H., Jones, C. J., & Wozniak, D. J. (2015). Bacterial extracellular polysaccharides in biofilm formation and function. *Microbial Biofilms*, 223-247.

Lira, M. C., Rodrigues, J. B., Almeida, E. T. d. C., Ritter, A. C., Tondo, E., Torres, S. M., Schaffner, D., de Souza, E. L., & Magnani, M. (2020). Efficacy of oregano and rosemary essential oils to affect morphology and membrane functions of noncultivable sessile cells of *Salmonella Enteritidis* in biofilms formed on stainless steel. *Journal of Applied Microbiology, 128*(2), 376-386.

Liu, F., Jin, P., Sun, Z., Du, L., Wang, D., Zhao, T., & Doyle, M. P. (2021). Carvacrol oil inhibits biofilm formation and exopolysaccharide production of *Enterobacter cloacae. Food Control, 119*, 107473.

Mao, Y., Liu, P., Chen, H., Wang, Y., Li, C., & Wang, Q. (2023). Baicalein Inhibits the *Staphylococcus aureus* Biofilm and the LuxS/AI-2 System in vitro. *Infection and Drug Resistance*, 2861-2882.

Mastoor, S., Nazim, F., Rizwan-ul-Hasan, S., Ahmed, K., Khan, S., Ali, S. N., & Abidi, S. H. (2022). Analysis of the antimicrobial and anti-biofilm activity of natural compounds and their analogues against *Staphylococcus aureus* isolates. *Molecules, 27*(20), 6874.

Minich, A., Levarski, Z., Mikulášová, M., Straka, M., Liptáková, A., & Stuchlík, S. (2022). Complex analysis of vanillin and syringic acid as natural antimicrobial agents against *Staphylococcus epidermidis* biofilms. *International journal of molecular sciences, 23*(3), 1816.

Mitra, S., Sultana, S. A., Prova, S. R., Uddin, T. M., Islam, F., Das, R., Nainu, F., Sartini, S., Chidambaram, K., & Alhumaydhi, F. A. (2022). Investigating forthcoming strategies to tackle deadly superbugs: current status and future vision. *Expert Review of Anti-infective Therapy, 20*(10), 1309-1332.

Morgan, J. L., McNamara, J. T., Fischer, M., Rich, J., Chen, H.-M., Withers, S. G., & Zimmer, J. (2016). Observing cellulose biosynthesis and membrane translocation in crystallo. *Nature, 531*(7594), 329-334.

Ortega-Ramirez, L. A., Gutiérrez-Pacheco, M. M., López-Romero, J. C., & Ayala-Zavala, J. F. (2021). Inhibition of Extracellular Polymeric Substances to Reduce Bacteria Pathogenicity. In K. D. W (Ed.), *Pathogenic Bacteria: Pathogenesis, Virulence Factors and Antibacterial Treatment Strategies* (pp. 1-56): Nova Science.

Ortega-Ramirez, L. A., Gutiérrez-Pacheco, M. M., Vargas-Arispuro, I., González-Aguilar, G. A., Martínez-Téllez, M. A., & Ayala-Zavala, J. F. (2020). Inhibition of glucosyltransferase activity and glucan production as an antibiofilm mechanism of lemongrass essential oil against Escherichia coli O157: H7. *Antibiotics, 9*(3), 102.

Packiavathy, I. A. S. V., Agilandeswari, P., Musthafa, K. S., Pandian, S. K., & Ravi, A. V. (2012). Antibiofilm and quorum sensing inhibitory potential of *Cuminum cyminum* and its secondary metabolite methyl eugenol against Gram negative bacterial pathogens. *Food Research International, 45*(1), 85-92.

Palomares-Navarro, J. J., Bernal-Mercado, A. T., González-Aguilar, G. A., Ortega-Ramirez, L. A., Martínez-Téllez, M. A., & Ayala-Zavala, J. F. (2022). Antibiofilm action of plant terpenes in *Salmonella* strains: Potential inhibitors of the synthesis of extracellular polymeric substances. *Pathogens, 12*(1), 35.

Pandit, S., Fazilati, M., Gaska, K., Derouiche, A., Nypelö, T., Mijakovic, I., & Kádár, R. (2020). The exo-polysaccharide component of extracellular matrix is essential for the viscoelastic properties of *Bacillus subtilis* biofilms. *International journal of molecular sciences, 21*(18), 6755.

Pantsar, T., & Poso, A. (2018). Binding affinity via docking: fact and fiction. *Molecules, 23*(8), 1899.

Passos da Silva, D., Matwichuk, M. L., Townsend, D. O., Reichhardt, C., Lamba, D., Wozniak, D. J., & Parsek, M. R. (2019). The *Pseudomonas aeruginosa* lectin LecB binds to the exopolysaccharide Psl and stabilizes the biofilm matrix. *Nature communications, 10*(1), 2183.

Patil, A., Banerji, R., Kanojiya, P., & Saroj, S. D. (2021). Foodborne ESKAPE biofilms and antimicrobial resistance: Lessons learned from clinical isolates. *Pathogens and Global Health, 115*(6), 339-356.

Pejin, B., Ciric, A., Glamoclija, J., Nikolic, M., & Sokovic, M. (2015). In vitro anti-quorum sensing activity of phytol. *Natural product research, 29*(4), 374-377.

Phillips, J. C., Braun, R., Wang, W., Gumbart, J., Tajkhorshid, E., Villa, E., Chipot, C., Skeel, R. D., Kale, L., & Schulten, K. (2005). Scalable molecular dynamics with NAMD. *Journal of computational chemistry, 26*(16), 1781-1802.

Ray, R. R., & Pattnaik, S. (2023). Contribution of phytoextracts in challenging the biofilms of pathogenic bacteria. *Biocatalysis and Agricultural Biotechnology, 48*, 102642.

Ren, Q., Luo, W., Chi, H., Zhang, L., & Chen, W. (2023). Down-regulation of β-lactam antibiotics resistance and biofilm formation by *Staphylococcus epidermidis* is associated with isookanin. *Frontiers in cellular and infection microbiology, 13*, 1139796.

Reza, A., Sutton, J. M., & Rahman, K. M. (2019). Effectiveness of efflux pump inhibitors as biofilm disruptors and resistance breakers in gram-negative (ESKAPEE) bacteria. *Antibiotics, 8*(4), 229.

Roux, D., Cywes-Bentley, C., Zhang, Y.-F., Pons, S., Konkol, M., Kearns, D. B., Little, D. J., Howell, P. L., Skurnik, D., & Pier, G. B. (2015). Identification of poly-N-acetylglucosamine as a major polysaccharide component of the *Bacillus subtilis* biofilm matrix. *Journal of Biological Chemistry, 290*(31), 19261-19272.

Sakarikou, C., Kostoglou, D., Simões, M., & Giaouris, E. (2020). Exploitation of plant extracts and phytochemicals against resistant Salmonella spp. in biofilms. *Food Research International, 128*, 108806.

Salmaso, V., & Moro, S. (2018). Bridging molecular docking to molecular dynamics in exploring ligand-protein recognition process: An overview. *Frontiers in pharmacology, 9*, 923.

Schneider-Rayman, M., Steinberg, D., Sionov, R. V., Friedman, M., & Shalish, M. (2021). Effect of epigallocatechin gallate on dental biofilm of *Streptococcus mutans*: An in vitro study. *BMC Oral Health, 21*(1), 1-11.

Schulze, A., Mitterer, F., Pombo, J. P., & Schild, S. (2021). Biofilms by bacterial human pathogens: Clinical relevance-development, composition and regulation-therapeutical strategies. *Microbial Cell, 8*(2), 28.

Seper, A., Fengler, V. H., Roier, S., Wolinski, H., Kohlwein, S. D., Bishop, A. L., Camilli, A., Reidl, J., & Schild, S. (2011). Extracellular nucleases and extracellular DNA play important roles in *Vibrio cholerae* biofilm formation. *Molecular microbiology, 82*(4), 1015-1037.

Shukla, A., Shukla, G., Parmar, P., Patel, B., Goswami, D., & Saraf, M. (2021). Exemplifying the next generation of antibiotic susceptibility intensifiers of phytochemicals by LasR-mediated quorum sensing inhibition. *Scientific Reports, 11*(1), 22421.

Siddharth, T., Sridhar, P., Vinila, V., & Tyagi, R. (2021). Environmental applications of microbial extracellular polymeric substance (EPS): A review. *Journal of Environmental Management, 287*, 112307.

Silva-Espinoza, B. A., Palomares-Navarro, J. J., Tapia-Rodriguez, M. R., Cruz-Valenzuela, M. R., González-Aguilar, G. A., Silva-Campa, E., Pedroza-Montero, M., Almeida-Lopes, M., Miranda, R., & Ayala-Zavala, J. F. (2020). Combination of ultraviolet light-C and clove essential oil to inactivate Salmonella Typhimurium biofilms on stainless steel. *Journal of Food Safety, 40*(3), e12788.

Silva, E., Teixeira, J. A., Pereira, M. O., Rocha, C. M., & Sousa, A. M. (2023). Evolving biofilm inhibition and eradication in clinical settings through plant-based antibiofilm agents. *Phytomedicine*, 154973.

Singha, S., Thomas, R., Kumar, A., Bharadwaj, D., Vishwakarma, J. N., & Gupta, V. K. (2023). Presence of potent inhibitors of bacterial biofilm associated proteins is the key to Citrus limon's antibiofilm activity against pathogenic *Escherichia coli. Biofouling, 39*(2), 171-188.

Smith, D. R., Price, J. E., Burby, P. E., Blanco, L. P., Chamberlain, J., & Chapman, M. R. (2017). The production of curli amyloid fibers is deeply integrated into the biology of *Escherichia coli. Biomolecules, 7*(4), 75.

Song, Y. J., Yu, H. H., Kim, Y. J., Lee, N.-K., & Paik, H.-D. (2019). Anti-biofilm activity of grapefruit seed extract against *Staphylococcus aureus* and *Escherichia coli.*

Srinivasan, R., Mohankumar, R., Kannappan, A., Karthick Raja, V., Archunan, G., Karutha Pandian, S., Ruckmani, K., & Veera Ravi, A. (2017). Exploring the anti-quorum sensing and antibiofilm efficacy of phytol against *Serratia marcescens* associated acute pyelonephritis infection in Wistar rats. *Frontiers in cellular and infection microbiology, 7*, 498.

Thakur, S., Ray, S., Jhunjhunwala, S., & Nandi, D. (2020). Insights into coumarin-mediated inhibition of biofilm formation in *Salmonella Typhimurium. Biofouling, 36*(4), 479-491.

Valdés-Tresanco, M. S., Valdés-Tresanco, M. E., Valiente, P. A., & Moreno, E. (2021). gmx_MMPBSA: a new tool to perform end-state free energy calculations with GROMACS. *Journal of chemical theory and computation, 17*(10), 6281-6291.

Van Der Spoel, D., Lindahl, E., Hess, B., Groenhof, G., Mark, A. E., & Berendsen, H. J. (2005). GROMACS: fast, flexible, and free. *Journal of computational chemistry, 26*(16), 1701-1718.

Varga, J. J., Therit, B., & Melville, S. B. (2008). Type IV pili and the CcpA protein are needed for maximal biofilm formation by the gram-positive anaerobic pathogen *Clostridium perfringens. Infection & immunity, 76*(11), 4944-4951.

Vazquez-Armenta, F., Bernal-Mercado, A., Tapia-Rodriguez, M., Gonzalez-Aguilar, G., Lopez-Zavala, A., Martinez-Tellez, M., Hernandez-Oñate, M., & Ayala-Zavala, J. (2018). Quercetin reduces adhesion and inhibits biofilm development by *Listeria monocytogenes* by reducing the amount of extracellular proteins. *Food Control, 90*, 266-273.

Wang, R., Vega, P., Xu, Y., Chen, C. Y., & Irudayaraj, J. (2018). Exploring the anti-quorum sensing activity of ad limonene nanoemulsion for *Escherichia coli* O157: H7. *Journal of Biomedical Materials Research, 106*(7), 1979-1986.

Wilbon, P. A., Chu, F., & Tang, C. (2013). Progress in renewable polymers from natural terpenes, terpenoids, and rosin. *Macromolecular rapid communications, 34*(1), 8-37.

Wu, Y.-P., Liu, X.-Y., Bai, J.-R., Xie, H.-C., Ye, S.-L., Zhong, K., Huang, Y.-N., & Gao, H. (2019). Inhibitory effect of a natural phenolic compound, 3-p-trans-coumaroyl-2-hydroxyquinic acid against the attachment phase of biofilm formation of *Staphylococcus aureus* through targeting sortase A. *RSC advances, 9*(56), 32453-32461.

Yong, Y. Y., Dykes, G. A., & Choo, W. S. (2019). Biofilm formation by staphylococci in health-related environments and recent reports on their control using natural compounds. *Critical reviews in microbiology, 45*(2), 201-222.

Zhang, C., Li, C., Abdel-Samie, M. A., Cui, H., & Lin, L. (2020). Unraveling the inhibitory mechanism of clove essential oil against Listeria monocytogenes biofilm and applying it to vegetable surfaces. *LWT, 134*, 110210.

Zhang, Y., Ding, Z., Shao, C., & Xie, J. (2022). Chlorophyllin-Based 405 nm Light Photodynamic Improved Fresh-Cut Pakchoi Quality at Postharvest and Inhibited the Formation of Biofilm. *Foods, 11*(16), 2541.

Zhou, L., Xia, S., Zhang, Z., Ye, B., Xu, X., Gu, Z., & Wang, X. (2014). Effects of pH, Temperature and Salinity on Extracellular Polymeric Substances of *Pseudomonas aeruginosa* Biofilm with N-(3-Oxooxtanoyl)-L-Homoserine Lactone Addition. *Journal of Water Sustainability, 4*(2), 91.

Chapter 4

Host Cell Proteins Enrichment for an In-Depth Analytical Assessment of Biopharmaceutical Quality

Egisto Boschetti[1]
and Pier Giorgio Righetti[2]
[1]Scientific Consultant, JAM Conseil, Neuilly-sur-Seine, France
[2]Department of Chemistry, Politecnico di Milano, Milan, Italy

Abstract

The question of HCP (host cell proteins) presence in recombinant therapeutic proteins is a sensitive subject impacting the stability and the safety of these biopharmaceuticals. Hence it is essential to devise reliable and very sensitive assays to identify and quantify these protein impurities at levels where the most common technologies are insufficient or even failing. Traditionally immunoenzymatic assays (ELISA) have been used; however, they suffer from limited sensitivity because some HCP are present at extremely low concentrations. Moreover, ELISA assays are unable to identify them. These weaknesses do not ensure the full elimination of HCPs, whose presence may impact the stability of the drugs and may introduce a risk of health to the receiving patients due to, for instance, potential immunogenic reactions. Recent efforts are focused on the use of mass spectrometry with the capability to individually identify HCPs and to quantify them. Nevertheless, very low-abundance host cell proteins escape detection because of the sample very large dynamic range. This situation promotes efforts around the enrichment of low-abundance HCP to improve the sensitivity of analytical methods. The present report is a short review of such an approach with a special focus on the use of combinatorial peptide ligand libraries (CPLL), extensively applied in proteomics investigations, for the enrichment of

In: Advances in Biology. Volume 5
Editor: Charles D. Grant
ISBN: 979-8-89113-441-6

very low-abundance proteins in view of their identification and individual quantitation.

Keywords: host cell proteins, proteomics, impurity identification, recombinant proteins, peptide library

Introduction

Biopharmaceuticals intended for human therapeutic use are most of the time proteins that are extracted from biological fluids such as human blood or are produced by recombinant technologies by *in vitro* cell culture processes. This latter approach is probably one of the most dynamic topics on bipharma industry today especially with the large-scale production of the wide diversity of therapeutic monoclonal antibodies [1]. Beyond antibodies, several other recombinant proteins such as vaccines, hormones, coagulation factors are also produced by cell culture. Hundreds of products can be counted that take their origin from the culture of recombinant procaryotic or eukaryotic cells. At present, however, most of these proteins are produced in mammalian expression systems because the ability of these cells to correctly produce proper post translational modifications (e.g., glycosylation) and correct protein folding [2]. The most popular expression system is based on recombinant CHO (Chinese hamster ovary) cells [3, 4] because they are capable of a relatively high levels of expression with the applicability at large scale in stable conditions and consistency. The recombinant biologicals that are harvested from cell culture are not pure and need a series of steps for their isolation to reach very high levels of purity. Impurities to be eliminated come from cell culture media and from the normal expression of the living cells as well as the lysis of cells during their life cycle. In this respect the downstream processing must be designed to reach a purity level required for human therapeutic applications ensuring safety and consistency. Some impurities related to the production process are difficult to eliminate since they can closely follow the target recombinant protein and thus co-purify with the final product although in much smaller quantities [5]. This is a class of protein impurities of great heterogeneity involving various risks for patients. Immunogenic risks [6], enzymatic risks [7] and proteolytic activity risks [8] are the most representative. Moreover, the potential to inhibit certain biological activities necessary for the patient who will receive the final biopharmaceutical can also be considered as a relevant risk. The most visible consequence of the effect of impurities is the progressive degradation of the finished product with a reduction in its biological activity for which it is produced. Because of these problems HCPs are particularly tracked during the purification process of the recombinant biopharmaceutical.

Although dedicated downstream purification processes can remove the HCPs, their traces could persist in the final drug product (around 100 ppm or less). Specific polishing steps have been designed over the years [9]; however, their efficacy may not always be optimal. This is why there is a need for HCP detection and quantification at different clearance stages [10, 11]. The procedure is practically mandatory but not yet highly standardized because HCP impurities largely depend on the type of cell selected for the production of the target protein and the genetic construction.

It is during the development of the purification plan which the choice of separation means (chromatographic supports) that are useful to better discriminate between HCP and the recombinant protein to be isolated must be made. The general purification principle is based on the capture of the main protein at the first purification step to eliminate the greatest number of contaminants [12] followed by other orthogonal separation phases. During the purification steps, HCP analyses must be carried out in order to determine the best operating conditions for deleting impurities. This is often done with immunohistochemical ELISA [13] assays. Although possessing undeniable and recognized global quantitative properties, these analyses lack complete coverage for all impurities (see below), which leaves other techniques, such as mass spectrometry, to manage the identification of each protein impurity as well as the quantification of each constituent [10, 14]. Mass spectrometry can be used alone or in conjunction with an enrichment strategy using for example protein A affinity chromatography [15]. By a similar way an effective enrichment of low-abundance proteins concomitant to the reduction of dynamic concentration range can be performed with the use of combinatorial libraries of peptides (CPLL) [16].

Detection of impurities present in the final product and total elimination are two different and complementary topics that our group examined more than 15 years ago. These two subjects are fully linked together but require different techniques. On the one hand, it is essential to look for the slightest impurities that accompany the therapeutic target protein with techniques to visualize them indiscriminately and then develop appropriate elimination processes. Since there are traces of proteins of very low-abundance, our research team had suggested using a performant enrichment technique based on CPLLs [17-19] more than a decade ago. This technique, forgotten for several years, seems becoming more and more popular. It is the purpose of this review to describe the developments of this technology in the field of HCP detection.

In this targeted review the current situation of analytical methods for HCP is examined, followed by technologies of HCP enrichment with a special focus on the use of combinatorial peptide ligand libraries. Relevant recent data with specific documented examples are given.

The Threats of HCP for Biopharmaceuticals

HCPs are complex protein impurities commonly associated with the production processes of biopharmaceutical and naturally originated by the expression systems [20]. The composition and quantity of HCPs are largely variable and dependent on several factors such as the culture conditions and the genomic constructs. The purification of the expressed recombinant biodrugs requires several fractionation steps involving diverse technical tools like filtration and chromatography in view of minimizing the presence of HCP associated with the recombinant injectable protein. This subject is so sensitive that many research centers devote specific efforts to understand, explain and propose rational solutions to better control the situation, as evidenced by the number of scientific publications that have been continuously increasing for several years (see Figure 1).

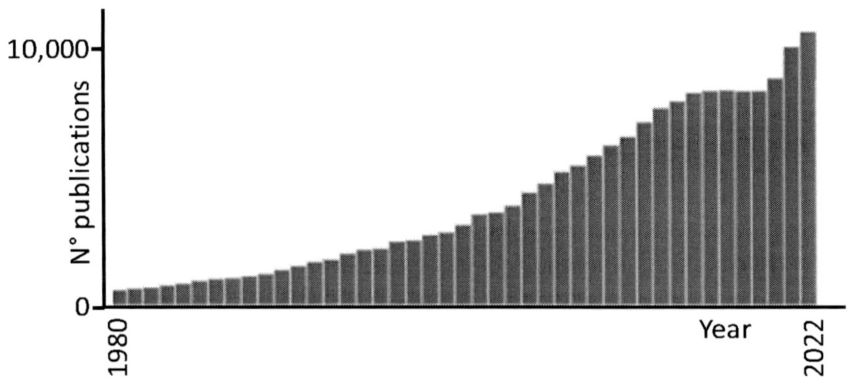

Figure 1. Progression of the number of HCP-based peer-reviewed publications over the last 40 years. It is interesting to note that even in the last few years the progression is still very high reaching an average of about 10% per year with more than 11,000 published papers for 2022. Data from PubMed (NIH, National Library of Medicine).

Although it is generally admitted that below 100 ppm the amount of host cell proteins is an acceptable level, these contaminating compounds could be present in large number and represent a threat in terms of final product stability and in terms of immunogenicity.

All protein impurities come from the recombinant cells that produce their own proteins. Other proteins come from the cell lysis and are not necessarily all secreted during the process of cell culture, as reported by Tait et al. [21]. Instead, they can come from the cell death and equally contribute to the contamination of expressed biopharmaceuticals.

As a reminder the most common host cells used for the production of recombinant proteins include *E. coli*, a bacterium with more than 4,000 genes [22], and Chinese Hamster Ovary (CHO) cells, a mammalian cell with about 20,000-25,000 genes [23]. Thus, the number of proteins constituting these cells and potentially present in the crude extracts of recombinant biopharma products is extremely large. Useless to say that the total elimination of these contaminants is very hard even with the most powerful purification processes especially for those that are qualified as protein impurities "difficult to remove". In an experimental work [24] it has been indicated that several HCP from CHO cells are of this category including cathepsin D, clusterin, galectin-3-binding protein, and few others. Hence these proteins appear as common in several recombinant proteins and of course in biosimilars suggesting dedicated downstream process design that could be generalized rendering "biosimilars more similar".

Proteins from host cells carry numerous biological activities such as hydrolytic enzymes (proteases and glycosidases); others carry immunogenic properties due to the phylogenic difference compared to the recipient patient treated with a contaminated biopharmaceutical.

The protease activity has been repeatedly described with the production of truncated forms and the presence of fragments of the target protein. Traces of protease activity can act on the long-term storage with a progressive degradation of the biodrug [18]. It is known in fact that recombinant cell cultures produce various lytic activities capable to progressively degrade the target recombinant proteins. Traces of proteolytic properties [25] and N-glycan degradation activities [11] that are detrimental to the long-term stability of the recombinant biodrugs are actually described. Residual proteolytic activity from the host cells still presents at the end of the purification process as for instance repeatedly described for monoclonal antibodies preparations [26] represent thus a threat for the long-term stability.

Peptides from proteolytic activity associated to other HCP traces can equally be detrimental to the patient health. In the presence of exogenous proteins, the human body may react with induction of unexpected immune responses that can result in serious critical situations [6, 27, 28] with possible clinical sequelae that are often challenging to predict.

The information about the immunogenic properties of certain HCP is not new, since it has been revealed decades ago upon injection of biodrugs [6, 29-31].

Another important point to consider is the question of "biosimilars". Although the target protein may be the same, coming from the same expression system, with similar degree of purity and biological activity, the downstream processing could be dissimilar leading to a variation of the diversity of HCP traces. Even if the quantity of HCPs is lower than the 100-ppm limit, the diversity and the nature of HCP might be a real threat. As said above among HCP there are some polypeptides that are difficult to remove and must be especially tracked in order to guide the process of downstream purification to reach a lower presence level [24]. The adverse reactions due to protein impurities are also contingent upon the type of therapeutic indication. There are pathologies implying just a single treatment and other such as hemophilia that suppose treatments for the lifetime.

To complete the picture, it should be said that in order to reduce the potential of immunogenic reactions and to better meet the right protein glycosylation, it is suggested that the best host cells could be of human origin [32]. Most of these arguments along with the limits of use of current analytical methods are discussed by Zhu-Shimoni et al. [10].

Current Methods for HCP Detection and Quantitation

Considered as components that may compromise the quality and safety of the final biopharmaceuticals, host cell proteins need to be investigated. The major obstacle is related to their presence in trace amount requiring very sensitive analytical methods. Although stringent purification procedures comprising special polishing treatments are intended to ensure the removal of foreign proteins [33, 34], to validate the efficacy of the purification process, dedicated and sensitive analytical assays have been devised.

Historically, the monitoring of HCPs has been a constant concern since the beginning of the development of biopharmaceuticals prepared by recombinant cell cultures [35, 36] and a quantitative limit of host protein impurities has been defined as 1-100 ng per mg of purified recombinant protein [37]; however, this limit was essentially dictated by the sensitivity of the analytical methods.

In a large study of the cell culture supernatant Zhang et al. [38] identified about 500 individual HCPs, but this number does not provide valuable information neither on the impact of each of them on the recombinant biopharmaceutical nor on the certitude of full detection of low-abundance HCPs. Therefore, the detection and the identification of HCP impurities is still considered as a real challenge in the production or recombinant proteins [36, 37].

Although the selection of expression systems is relatively large from procaryotic to eukaryotic cells rendering the investigations even more challenging, the choice of the host is more and more focused on mammalian Chinese Hamster Ovary (CHO) cells enhancing thus the level of knowledge due to increasingly larger scientific HCP documentation. Despite the availability of data, a particular attention has still to be paid to define acceptable level of protein impurities because, beyond the well-established rules (less than 100 ng/mL and no more than 10 ng per dose) [39], the question is largely more complicated.

An overview of the HCP profile and different analytical technologies is reported by Bracewell et al. [40].

A current method for the measurement of HCPs in recombinant pharmaceutical proteins all along the purification process and on the final purified product, is ELISA (see for review [10] and [41]). An HCP ELISA assay is based on the preparation and use of polyclonal antibodies obtained by immunizing animals such as sheep or goats with a host cell extract. Such an approach is not exempt of criticism because host proteins have not a similar immunogenicity and/or may generate antibodies of low affinity reducing thus their ability to detect HCPs properly. In this situation the detection may not cover the complete spectrum of HCPs within the recombinant product.

Anti-HCP antibodies have been equally applied to Western blotting of electrophoretic techniques with similar limitations due to the quality of the pool of antibodies [42]. Hence suggestions have been made to make correlations with orthogonal analytical determinations with for instance protein precipitation, mono and bi-dimensional polyacrylamide gel electrophoresis and mass spectrometry [21-43]. 2D-DIGE has also been used to compare the presence of HCP during the purification process of recombinant proteins [44]. Immuno-specific methods and non-specific approaches have thus been associated to better cover the detection of HCP.

The methods include ELISA, 2DE, immunoblots, 2D-DIGE and 2D-HPLC [45].

In a recent study the combination of ELISA based immunodetection with LC-MS/MS (liquid chromatography - tandem mass spectrometry) demonstrated an improved approach to accurately perform the measurement of the HCP coverage [46]. The authors expressed the hope that the methodology would be advantageously applied by biopharma industry to better monitor the purification process and consequently to improve the safety for patients.

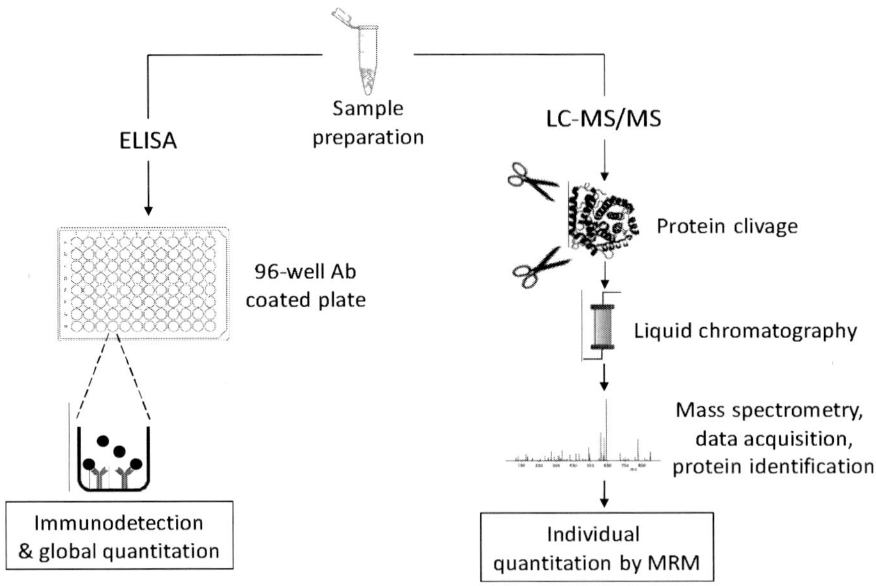

Figure 2. Experimental general setup for the detection and quantification of HCP. On the left it is represented the streamline of ELISA procedure with the detection in different samples on a 96-wells plate. When using a polyclonal antibody against the proteome of the host cells, the final results are expressed as a global amount of HCPs in each sample. On the right the process involving mass spectrometry is roughly represented. The initial sample is trypsinized first and resulting peptides are analyzed by mass spectrometry. All obtained data are then used to identify each HCP present in the sample. To make the quantitative determination it is common to use MRM technology applied to each identified HCP. In both cases the analysis can be addressed to detectable HCPs since all low-abundance species are at a concentration below the detection limits of both technologies. They can be reached only after having applied specific methods of protein enrichment.

A particular performant technology is mass spectrometry with all possible variations. Even at present mass spectrometry continues making progress when used alone or in combination with other above-described methods [47]. It can monitor multiple protein analytes at a time at a high throughput and allows identifying protein impurities while giving acceptable quantitative estimations. Mass spectrometry largely used in proteomics studies can easily be associated with chromatography, electrophoresis and various labeling methods enhancing the capabilities to track HCP. Nevertheless, all the above-described methods including mass spectrometry suffer from their inability to detect low-abundance proteins that are either below the sensitivity of the instruments or obscured by the signal of high abundance proteins. Although the performance of the instrumentation is improving over time, the detection of HCP traces remains challenging when in presence of the dominant expressed recombinant biopharmaceutical.

The Need of HCP Enrichment Prior Analysis

The detection and identification of HCPs in recombinant biodrugs is reminiscent of proteomic investigations where the track of very low-abundance protein is one of the most difficult subjects resulting from the extreme dynamic concentration range of protein components. Similarly, whatever the expression system, the dilemma of large dynamic concentration range of host proteins alone also remains a hard technical problem. Here the question is even harder since beyond the natural dynamic range of host cell proteins themselves, there is the presence of the extreme high-abundance recombinant protein (the biodrug) especially after the overall purification process. In effect HCP polypeptides abundance can be 4 to 10 orders of magnitude lower than the recombinant therapeutic protein. This massive presence naturally masks the signal of most if not all HCP of very low abundance in two-dimensional electrophoresis as well as in mass spectrometry. With this in mind, to resolve the detection of HCP several methods have been designed with the objective to reduce the concentration of the main protein as much as possible, while enriching all other protein impurities. Among the most known method are the so-called depletion of the recombinant protein, immunoprecipitation, limited digestion, cutoff filtration, size exclusion, various chromatographic procedures, and CPLL technology. Each of these procedures has its own advantages, but is not exempt of criticism as illustrated on Table 1.

Table 1. Comparative properties of the most known methods allowing the detection of low-abundance proteins such as HCP

Enrichment methods	Advantages	Limitations	References
Immunoprecipitation	Specific collection of HCP	Possible coprecipitation and HCP losses. Requires anti-host proteins antibodies. No collection of poorly immunogenic HCP.	[16, 80]
Depletion	Better visibility of LAP by elimination of the recombinant main protein.	Possible co-depletion. Requires anti main protein ligand. LAP still very dilute. No reduction of dynamic range of HCP.	[49, 50, 87]
Cutoff filtration & HILIC	Elimination of HAP with concomitant collection of LAP	Dilution effect. No reduction of dynamic range of HCP.	[51, 53]
Size exclusion	Separation of the main protein if of different molecular mass.	Large dilution effect requiring concentration. No reduction of dynamic range of HCP.	[52]
Combinatorial affinity peptides	Large reduction of the dynamic range. Applicable to any HCP enrichment	Needs large samples.	[16, 19, 80]

HCP: host cell proteins.
HAP: high-abundance proteins.
LAP: low-abundance proteins.
HILIC: hydrophilic interaction liquid chromatography

Immunodepletion of the recombinant protein is in appearance the most obvious way to follow. It removes specifically the main protein (the biopharma product) leaving all remaining host proteins available for analysis. However, several barriers are to overcome. The first is to have available the appropriate solid phase grafted with the specific antibody. A second obstacle is the low binding capacity of most immunosorbents so that only very small samples could be treated each time, hence very small amounts of HCPs are involved. Moreover, immunodepletion does not improve the very large dynamic range of HCP themselves where the low-abundance species remain most of the time undetectable. To all these limitations are the losses of certain HCPs by nonspecific binding on the solid phase affinity immunodepletion media like in proteomic investigations known as co-depletion [48]. In spite of these limitations, this approach is used when dealing with recombinant monoclonal antibodies because the most common selective sorbent is based on protein A which better qualifies this technique as protein A affinity depletion rather than immunodepletion [15, 49, 50]. Basically, the sample to deplete is put in contact with Protein A solid phase and the supernatant is ready for the determination of HCPs possibly after concentration of remaining protein components (the HCP).

When the molecular mass difference between the recombinant protein and the HCP is significant, the enhancement of the latter group of proteins become better detectable by using technologies based on the properties of molecular mass separations such as cutoff enrichment [51] and molecular sieving chromatography [52]. Both methods eliminate the recombinant proteins suppressing thus the major interference in detection, but have limited applications due to losses in the former case and a large dilution in the latter case. Other chromatographic-related technologies have also been described such as hydrophilic interaction chromatography [53], hydrophobic chromatography [54] and optimized chromatographic gradients after limited trypsin digestion [55].

The search for an optimized procedure for both decreasing drastically the presence of the recombinant protein and the enhancement of HCP with an additional enrichment of low-abundance HCP seems to date achievable using CPLL beads as described in the following sections.

General Information on CPLLs and their Use

The enrichment of low-abundance proteins by combinatorial peptide ligand library (CPLL) is based on three main technical pillars: a mixed bed of affinity solid phase ligands (the peptide library), the capture of all protein from a sample under large overloading conditions and the exhaustive recovery of captured proteins.

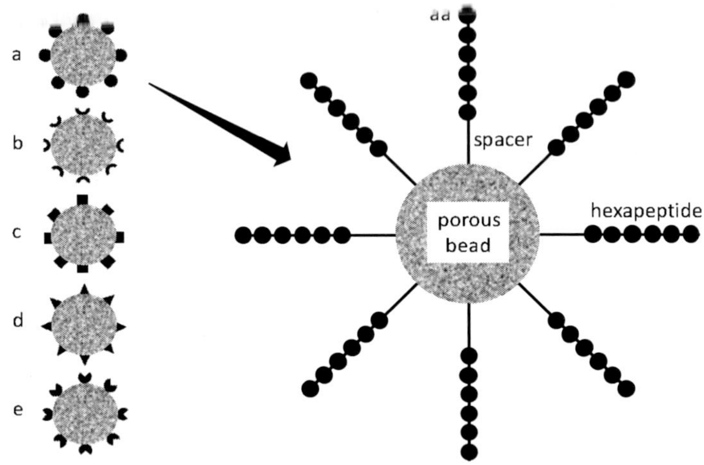

Figure 3. Scheme of the architecture of CPLL porous polymeric beads on which hexapeptide ligands are covalently attached (left panel; each bead with a different hexapeptide ligand). The bead structure consists of an organic porous polymer on whose matrix peptide ligands are covalently attached. On the right is a blow-up of a single bead around which numerous identical hexapeptides are attached by means of a spacer. "R" represents the chemical link. The peptide is attached *via* a spacer.

The solid phase is composed of porous microbeads each of them carrying many copies of a unique hexapeptide. Each bead carries a different peptide and thus acts as a single affinity chromatography sorbent (see scheme in Figure 3). As any molecular affinity-based mechanism, the interaction between the ligand and the proteins from the sample is managed by forces that depend on the amino acid composition of the peptide. Experimental studies [56, 57] showed that single peptides had the property to capture one or a group of proteins and all together, as a mixed bed, capture all proteins from a sample up to the bead's saturation. The diversity of ligands is a fundamental principle of this technology. A rational and complete explanation of this phenomenon has been reported a few years ago [58]. The major technical condition for the proper use of CPLL is thus the sample overloading which is mandatory for the compression of the dynamic concentration range consisting in the concentration of low-abundance proteins and the reduction of high-abundance ones.

Here as well, technical demonstrations have been reported for a variety of protein samples [59, 62]. Under overloading conditions, the dilute proteins accumulate progressively on their affinity beads while the most concentrated proteins saturate rapidly their corresponding beads. In addition to this behavior, an intense displacement effect takes place depending on the mass action law, which is a well-known phenomenon in current chromatographic sorbents as described for ion exchangers [63, 64] and hydrophobic sorbents [65]. This principle has already been used for the enrichment of certain categories of low-abundance proteins in lectin affinity chromatography [66] and other methods such as displacement chromatography [67]. It is thus the association of large affinity libraries of ligands and sample overloading conditions that constitute the concept of CPLL. To complete the picture, it should be mentioned that the capture of proteins is regulated by multiple operating parameters such as buffer ionic strength, temperature, pH and the presence of competitors.

Once the proteins capturing is completed, adsorbed proteins are harvested and identified by current analytical methods. To this end the full protein collection prior separation/identification is one option. The analysis is performed by electrophoresis (mono and two-dimensional) [68] and/or mass spectrometry [69, 70]. A variety of elution methods has been repeatedly reported [71, 73].

An alternative to protein elution is the direct on-bead trypsination followed by mass spectrometry also called Multidimensional Protein Identification Technology (MudPIT). Although less popular, this technique has the advantage to save working time and to allow identifying all proteins from a sample without complicated preliminary protein fractionation. Described in several circumstances this approach claims a larger number of detected low-abundance proteins compared to protein elution. This efficacy difference take origin from the number of diverse peptides generated by each protein treated with trypsin. In fact, while tightly adsorbed proteins can be difficult to elute, escaping thus detection, peptides are easy to collect and even if some of them are lost, others are still present to compensate this failure. In fact, the identification of one protein by peptide fingerprint is not contingent upon the collection of all peptides; most of the time two peptides are enough for a reliable identification.

This procedure of bead trypsination was applied in 2015 to the discovery of low-abundance protein of synovial fluid in osteoarthritis [74] with an enhanced number of proteins found exceeding 30% when compared to current elution methods. Few years later, the method was even improved with an optimized protocol showing an increased number of protein identifications and an increment in technical reproducibility. To reach this goal a pre-digestion treatment was adopted prior trypsination [75]. With these examples in mind, it is easy to conclude that the discovery of HCPs with a maximum efficacy would probably be performed under conditions capable to maximize the results while saving time especially when the method is applied to the optimization of the downstream processing.

Typical Examples of HCP Enrichment by Means of CPLLs

Analytical methods for the identification and quantitation of host cell proteins are hampered by the extremely wide dynamic protein concentration ranges between the main recombinant protein and the HCP traces. In addition to this real obstacle there is the fact that HCP proteins may change during the purification process. They can produce fragments and may be present under very different proportions. They are so diverse that the identification of single components is hard and constitutes a challenge to the quantitation of each of them. This is the reason why the HCPs are only roughly quantified "en mass" (all together) by polyclonal antibodies-based ELISA assays.

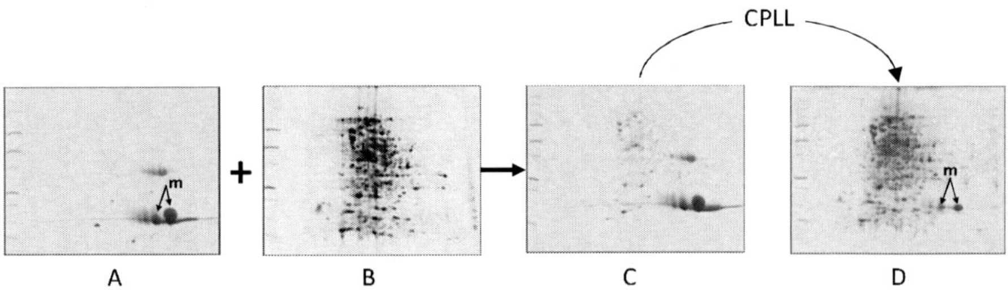

Figure 4. Two-dimensional electrophoresis analyses evidencing the HCP enrichment of a model sample by CPLL technology. 100 mg of pure myoglobin from sperm whale (panel A) was spiked with 5 mg of *E. coli* protein extract (panel B). The spiked sample representing a model of protein with procaryotic HCPs (panel C) was treated with CPLLs in view to enhance the presence of HCPs while reducing the content of myoglobin (panel D). The enrichment process was performed in 25 mM phosphate, pH 7.0: 400 μL of contaminated myoglobin solution were mixed with 80 μL of peptide library beads for 20 minutes at room temperature. After washing the proteins adsorbed on the beads were eluted using a denaturing buffer and analyzed. 2D-PAGE was performed in a pH gradient between 3 and 10 (first dimension) and using a 12% SDS-polyacrylamide in the second dimension. Protein staining by Coomassie blue.

As mentioned above, the use of CPLLs represents a very effective means to bring to the limelight low-abundant proteins by a special enrichment method based on multiple affinity with a large range of hexapeptides. In this section we illustrate the process by a series of experimental results covering a variety of situations.

Preliminary experiments were performed using a highly pure protein spiked with foreign proteins taken as a model for host cell proteins. Namely pure myoglobin was spiked with a soluble protein extract of *E. coli*. This mixture was analyzed as it was and after a treatment with CPLL. The obtained results are illustrated on Figure 4 where it is clearly demonstrated that the main protein (myoglobin) was largely reduced while traces of *E. coli* proteins concentrated in a single pass.

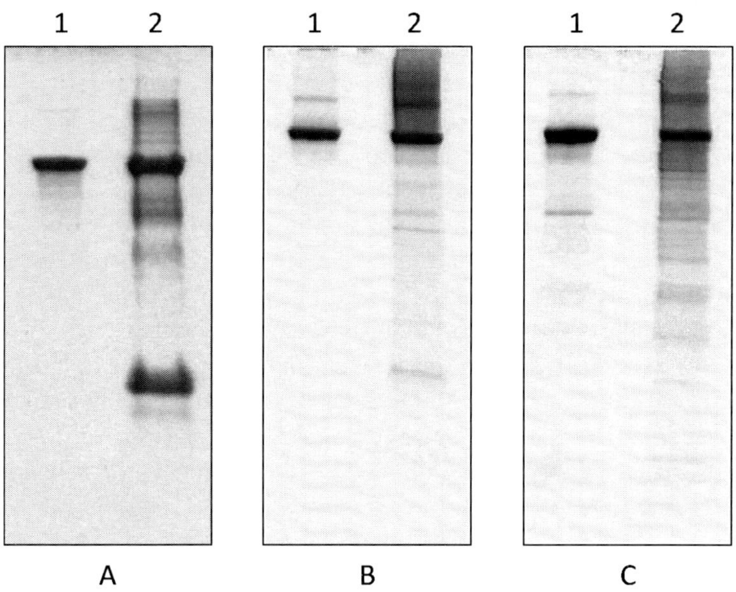

Figure 5. SDS-polyacrylamide gel electrophoresis of purified recombinant proteins where HCP traces have been enriched by a treatment with CPLLs. A: Native Rec-protein A classically purified by affinity chromatography on IgG grafted solid phase (lane 1) compared to the same sample after CPLL treatment (lane 2). Additional protein bands illustrate host cell proteins and fragments from proteolysis. B and C: Purified monoclonal antibodies respectively purified by liquid chromatography on MEP-Hypercel and Q-HyperD (lanes 1). Lane 2 represents the same samples after treatment with CPLL evidencing a multitude of host cell proteins and polypeptide fragments undetectable in the native mAbs.

A pioneering work on a real case has been made later when attempting to determine if highly purified recombinant Protein A expressed in genetically modified *E. coli* could be contaminated by protein traces from the host cells. Thus, several-grams of protein A have been submitted to a treatment of CPLL in view of amplifying the signal of possible traces of protein impurities [17]. What was found is a multitude of foreign proteins some of them being hydrolysis products of protein A and others from the host *E. coli* cells. The analyses were performed by SDS-polyacrylamide gel electrophoresis in combination with mass spectrometry. Essential results are illustrated on Figure 5-A. Among protein impurities, heat shock protein, riboflavin synthase beta chain and other numerous hypothetical proteins from *E. coli* were found. A similar work was performed two years later in view of tracking HCP from two purified recombinant monoclonal antibodies [18]. Here again host cell protein impurities were found along with proteolysis fragments from the main protein evidenced by SDS-polyacrylamide gel electrophoresis (see Figure 5-B), two-dimensional electrophoresis and mass spectrometry. Among HCPs were nucleoside diphosphate kinase B, peroxiredoxin-6 and -1, and alpha-2-macroglobulin precursor; other polypeptides were protein fragments, such as light and heavy chains. To complete this preliminary work, HCP detection and identification from purified recombinant human albumins expressed by recombinant *P. pastoris* was performed before and after enrichment of the low-abundance proteins by CPLL. The first analytical determination was made by two-dimensional electrophoresis. Separated protein spots were then individually excised, trypsinized and identified by mass spectrometry. As illustrated in Figure 6, the two-dimensional electrophoresis analysis of the purified recombinant albumin did not reveal significant protein impurities. However, when the sample was submitted to an enrichment process, numerous impurities were detectable. LC-MS/MS analysis confirmed the presence of HCPs such as imidazoleglycerol-phosphate dehydratase and ornithine carbamoyltransferase. Moreover, numerous polypeptides generated by proteolysis of albumin were identified within a large range of molecular masses and isoelectric points.

Despite these experimental unequivocal demonstrations, the enrichment of HCPs by CPLLs was not adopted for years. For commodity only ELISA-based analysis were extensively used for the quantitation of global HCPs although intrinsic limitations were well known (see previous chapter).

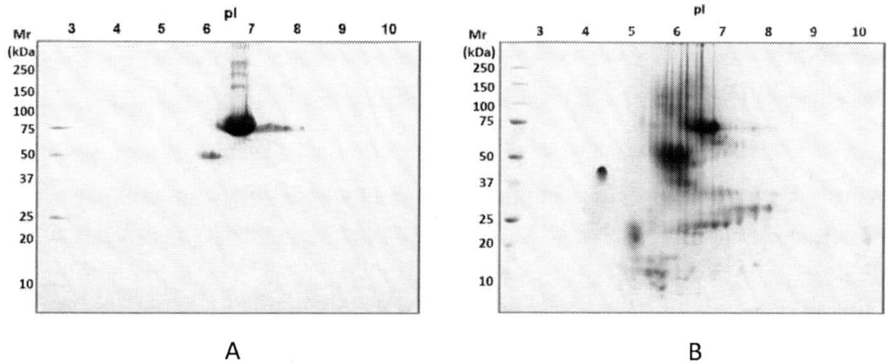

Figure 6. Two-dimensional electrophoresis analyses of purified recombinant human albumin before (A) and after treatment with CPLL (B) in view of enriching for low-abundance HCP. The separated protein spots are then excised, trypsinized and analyzed by mass spectrometry in view of individual protein identification.

It is only in the recent years that the CPLL approach for the detection of HCPs aroused a large scientific interest as witnessed by a growing number of published reports. In one of them Lavoie et al. [76] described further the technical concept and tried to compare a variety of combinatorial peptides (tetra and hexa) in order to identify those that were better designed for the capture of HCP while leaving the maximum amount of antibodies in the supernatant. In this work tetrameric multipolar peptides appeared as the most optimal for this application. The authors underlined the efficacy of such solid-phase affinity-like adsorbents for the elimination of impurities from the recombinant proteins. The direct consequence of this application work is naturally the enrichment of HCPs that could be collected as impurities present in the recombinant target protein and analyzed.

In 2020 Mortstedt et al. [77] tested CPLL beads to amplify the signal of HCPs. To this end a protein expressed by genetically modified CHO cells was spiked with six known proteins at concentrations ranging from 10 to 1000 ppm. The mixture was submitted to CPLL treatment and collected proteins were trypsinized and analyzed by LC-MS/MS against a control. To make an estimation of the abundance of spiked proteins an MRM (multiple reaction monitoring) analysis was applied. In addition to the six spiked standard proteins, 30 others host proteins were found. The former group of proteins was detected to 30 ppm in the CPLL-treated sample while no spiked proteins were detectable in the control. Concerning the HCP proteins, an MRM assay way devised for 14 of them. By that way it has been shown that the enrichment effect for HCP proteins was between 80 to 700 folds depending on the targeted protein.

Certain HCPs are particularly problematic for the stability of the therapeutic recombinant proteins such as lipase and carboxylesterase. For them Yen et al. [78] defined a practical strategy for their identification and quantitation. The workflow included either a direct digestion with trypsin followed by a LC-MS/MS-MRM assay reaching a sensitivity of 0.05 ppm in recombinant monoclonal antibodies. The method seems adapted to routine application during process development and is extendable to the detection and quantitation of other problematic HCPs and to other therapeutic protein drug candidates.

The detection of others HCPs in therapeutical monoclonal antibodies is one of the important subjects of investigation for recombinant proteins. Explanations of such a focus is due to the rapid growing of their design and large-scale production addressing increasingly more pathologies. In a recent study Molden et al. [79] examined the content in HCPs of 29 commercially available monoclonal antibodies intended for therapeutic applications approved by the FDA between 1997 and 2021. Using liquid chromatography associated to mass spectrometry they found 79 individual proteins from the host. The amount of these foreign proteins was at an average concentration of 20 ppm whereas in certain preparations the HCP amount was not higher than 7 ppm. As a conclusion of this large investigation, it has been asserted that the data could be used as general information for future productions of mAbs and maybe also for an updated design of ELISA assays. Nevertheless, the expression systems are diverse and the genomic constructs are complex suggesting possible modifications in the production of host proteins and their analysis adaption according to the overall expression system. To try simplifying the situation in terms of the nature of host cell proteins and for a more rational global quantitation of them along with a better and standardized protein production, the choice of expression system converges progressively towards Chinese hamster ovary cells [25]. However, this option does not resolve the questions related to non-immunogenic HCPs and their formal identification. As a contribution to better detect very low-abundance protein impurities, Chen et al. [80] proposed to associate an HCP enrichment step with ProteoMiner followed by mass spectrometry detection. In this study a demonstration was made that certain low-abundance HCPs were enriched up to 1000-fold and easily detected and identified. The experimental work comprised first a model with purified antibodies spiked with known amounts of protein traces that could be detected at 0.05 ppm. When applied to real cases of commercially available monoclonal antibodies 527 hamster proteins were identified in the CPLL eluate with high confidence; the enrichment factor range was estimated of about 100-400-fold. The enrichment process was not the same for each HCP but was always substantial. In the best cases, such as mitochondrial malate dehydrogenase, the enrichment was up to 1109-fold while in the worst cases it was still of few dozens. Triplicate experiments indicated that the technology was confidently reproducible without ambiguity and that the quantitation of each protein component using the label-free MRM approach for each peptide in all runs showed very little variations. The authors concluded that their suggested approach could not only be applied in the analytical determinations of the final biopharma product, but also be an effective support to process developments and the optimization of purification protocols all along the monoclonal antibody production.

Although an amount of HCPs below 100 ppm does not seem impacting the safety of biodrugs, it is apparently not the case for their long-term stability. A recent report described that very low content (below ppm) of lipases, elastases and proteases, can engender degradation of common stabilizing excipients such as polysorbate on the one hand [81-83] and produce partial fragmentation of the protein drug on the other hand [84].

In the most recent years a trend is identified towards the search of better technologies to resolve the identification, the quantitation and the understanding of the effects of contaminating proteins coming from engineered host cells. It is within this context that an ultrasensitive method for HCPs profiling involving CPLL enrichment has been devised as reported [16]. Here the practical task was to check for protein traces in recombinant monoclonal antibody preparations that can possibly be detrimental to drug quality. This study significantly improved the degree of enrichment by coupling the CPLL enrichment technology to a limited digestion with trypsin. This technology, already used in previous described works [80], was here further improved thanks to the association with a limited digestion under various combinations. The authors compared three options. First, the sample was treated with CPLLs and the captured low-abundance proteins recovered by elution and trypsinized. In a second option, a partial trypsination was directly performed on beads. The third approach was the recovery of captured proteins and submission to a partial trypsination. This latter procedure allowed producing significantly better results as compared to the control and to more classical approaches (Figure 7).

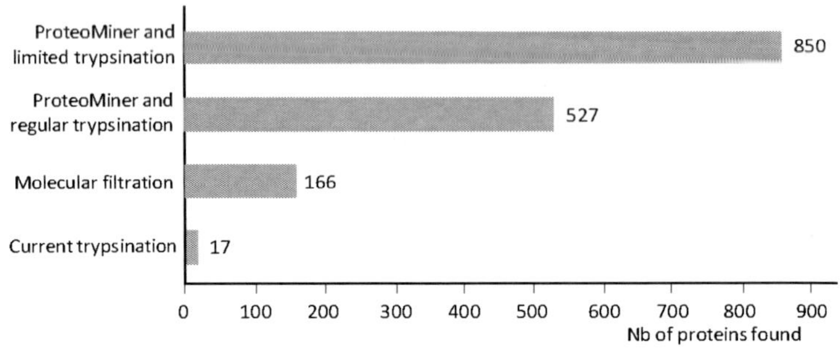

Figure 7. Comparative effect produced by various enrichment methods to detect low-abundance HCP in a purified recombinant monoclonal antibody. Data from Zhang et al. [16].

Upon mass spectrometry the number of proteins identified using two or more peptides was in the best case 850 with a false positive rate lower than 0.01. The reproducibility of the method was also evaluated. Three experiments were performed showing 759 common proteins or 80.1% of all proteins. The evaluation of the method on monoclonal antibodies demonstrated that low abundance HCP were enhanced 7694-fold with the capability to detect around 0.05 ppm with a best result for spiked thioredoxin of 0.002 ppm. With 850 host cell proteins found, this new method roughly represents an improvement of about 40% compared to what was previously described. The reduction of dynamic concentration range was estimated of about two orders of magnitude over more classical enrichment methods.

Conclusion

It is clear that CPLLs allow an amplification-like of foreign proteins coming from the host cells during the recombinant production of biopharma drugs. With this technology it becomes possible to better evaluate the amount of these proteins and identify each of them. The analysis of the CPLL-derived sample such as two-dimensional electrophoresis and mass spectrometry, drives the identification of single HCP component with the possibility of preparing the conditions of single quantitation. This can be easily performed by producing specific antibodies or by devising dedicated MRMs. When these options will become current reality, the follow-up of each downstream processing step will permit to better design the purification procedures of biodrugs. Certain host proteins such as those classified as "hard to remove" could equally be targeted when they represent a real threat for the patient or for the stability of the final drug. At this stage immunoblots could be redesigned as well as ELISA-based assays with probably a better co-involvement of both immunoassays and mass spectrometry procedures.

Advanced cell engineering approaches may allow producing modifications of the behavior of the host cells to reduce the presence of protease-like HCPs while improving the quality of the recombinant biodrugs [25, 85].

Another interesting center of interest, where the adoption of advanced analytical technologies can be applied, is the domain of "biosimilars" that are considered equivalent drugs but are not necessarily conceived on the same basis and hence could carry different HCPs. This question has been very recently discussed by Gibofsky et al. [86] where a literature survey over the last few years was made. Biosimilars, as for example monoclonal antibodies, are making progress in terms of availability; however, in spite of potential cost benefits and improved biomanufacturing, they show acceptability concerns. Interchangeability and HCP identity especially those that are very dilute, are questionable and are at the basis of reluctance to accept these drugs by the patients; moreover, they are administered with caution by practitioners. When the identification of each low-abundance HCP will be of general application, a better understanding of their potential side effects will be elucidated and a larger utilization of biosimilars should be adopted. In this endeavor the enrichment of low- and very low-abundance host cell protein contaminants would become more and more central and critical.

References

[1] Kaplon, H., Muralidharan, M., Schneider, Z., and Reichert, J. M. (2020). Antibodies to watch in 2020. *Mabs*, 12, 1703531.

[2] Kuriakose, A., Chirmule, N., and Nair, P. (2016). Immunogenicity of biotherapeutics: Causes and association with posttranslational modifications. *J Immunol Res*, 2016, 1298473.

[3] Lalonde, M. E., and Durocher, Y. (2017). Therapeutic glycoprotein production in mammalian cells. *J Biotechnol*, 251, 128–40.

[4] Tihanyi, B., and Nyitray, L. (2020). Recent advances in CHO cell line development for recombinant protein production. *Drug Discov Today Technol*, 38, 25-34.

[5] Vanderlaan, M., Zhu-Shimoni, J., Lin, S., Gunawan, F., Waerner, T., and Van Cott, K. E. (2018). Experience with host cell protein impurities in biopharmaceuticals. *Biotechnol Prog*, 34, 828-37.

[6] Jawa, V., Joubert, M. K., Zhang, Q., Deshpande, M., Hapuarachchi, S., Hall, M. P., and Flynn, G. C. (2016). Evaluating immunogenicity risk due to host cell protein impurities in antibody-based biotherapeutics. *AAPS J*, 18, 1439-52.

[7] Li, X., Chandra, D., Letarte, S., Adam, G. C., Welch, J., Yang, R. S., Rivera, S., Bodea, S., Dow, A., Chi, A., Strulson, C. A., and Richardson, D. D. (2021). Profiling active enzymes for polysorbate degradation in biotherapeutics by activity-based protein profiling. *Anal Chem*, 93, 8161-9.

[8] Cui, T., Chi, B., Heidbrink, Thompson, J., Kasali, T., Sellick, C., Turner, R., and Cathepsin, D. (2019). Removal strategy on protein A chromatography, near real time monitoring and characterization during monoclonal antibody production. *J Biotechnol*, 305, 51-60.

[9] Huang, Y., Molden, R., Hu, M., Qiu, H., and Li, N. (2021). Toward unbiased identification and comparative quantification of host cell protein impurities by automated iterative LC-MS/MS (HCP-AIMS) for therapeutic protein development. *J Pharm Biomed Anal*, 200, 114069.

[10] Zhu-Shimoni, J., Yu, C., Nishihara, J., Wong, R. M., Gunawan, F., Lin, M., Krawitz, D., Liu, P., Sandoval, W., and Vanderlaan, M. (2014). Host cell protein testing by ELISAs and the use of orthogonal methods. *Biotechnol Bioeng*, 111, 2367-79.

[11] Li, X., An, Y., Liao, J., Xiao, L., Swanson, M., Martinez-, Fonts, K., Pavon, J. A., Sherer, E. C., Jawa, V., Wang, F., Gao, X., Letarte, S., and Richardson, D. D. (2021). Identification and characterization of a residual host cell protein hexosaminidase B associated with N-glycan degradation during the stability study of a therapeutic recombinant monoclonal antibody product. *Biotechnol Prog*, 37, e3128.

[12] Jones, M., Palackal, N., Wang, F., Gaza-Bulseco, G., Hurkmans, K., Zhao, Y., Chitikila, C., Clavier, S., Liu, S., Menesale, E., Schonenbach, N. S., Sharma, S., Valax, P., Waerner, T., Zhang, L., and Connolly, T. (2021). "High-risk" host cell proteins (HCPs): A multi-company collaborative view. *Biotechnol Bioeng*, 118, 2870-85.

[13] Gunawan, F., Nishihara, J., Liu, P., Sandoval, W., Vanderlaan, M., Zhang, H., and Krawitz, D. (2018). Comparison of platform host cell protein ELISA to process-specific host cell protein ELISA. *Biotechnol Bioeng*, 115, 382-9.

[14] Waldera-Lupa, D. M., Jasper, Y., Köhne, P., Schwichtenhövel, R., Falkenberg, H., Flad, T., Happersberger, P., Reisinger, B., Dehghani, A., Moussa, R., and Waerner, T. (2021). Host cell protein detection gap risk mitigation: quantitative IAC-MS for ELISA antibody reagent coverage determination. *Mabs*, 13, 1955432.

[15] Thompson, J. H., Chung, W. K., Zhu, M., Tie, L., Lu, Y., Aboulaich, N., Strouse, R., and Mo, W. D. (2014). Improved detection of host cell proteins (HCPs) in a mammalian cell-derived antibody drug using liquid chromatography/mass spectrometry in conjunction with an HCP-enrichment strategy. *Rapid Commun Mass Spectrom*, 28, 855-60.

[16] Zhang, S., Xiao, H., and Li, N. (2022). Ultrasensitive method for profiling host cell proteins by coupling limited digestion to ProteoMiner technology. *Anal Biochem*, 657, 114901.

[17] Fortis, F., Guerrier, G., Areces, L., Antonioli, P., Hayes, T., Carrick, K., Hammond, D., Boschetti, E., and Righetti, P. G. (2006). A new approach for the detection and

identification of protein impurities using combinatorial solid phase ligand libraries. *J Proteome Res*, 5, 2577-85.

[18] Antonioli, P., Fortis, F., Guerrier, L., Rinalducci, S., Zolla, L., Righetti, P. G., and Boschetti, E. (2007). Capturing and amplifying impurities from purified recombinant monoclonal antibodies via peptide library beads: a proteomic study. *Proteomics*, 7, 1624-33.

[19] Righetti, P. G., Boschetti, E., and Fasoli, E. (2011). Capturing and amplifying impurities from recombinant therapeutic proteins via combinatorial peptide libraries: a proteomic approach. *Curr Phar Biotech*, 12, 1537-47.

[20] Wang, X., Hunter, A. K., and Mozier, N. M. (2009). Host cell proteins in biologics development: Identification, quantitation and risk assessment. *Biotechnol Bioeng*, 103, 446–58.

[21] Tait, A. S., Hogwood, C. E., Smales, C. M., and Bracewell, D. G. (2012). Host cell protein dynamics in the supernatant of a mAb producing CHO cell line. *Biotechnol Bioeng*, 109, 971-82.

[22] Blattner, F. R., Plunkett, G., III., Bloch, C. A., Perna, N. T., Burland, V., Riley, M., Collado-Vides, J., Glasner, J. D., Rode, C. K., Mayhew, G. F., Gregor, J., Davis, N. W., Kirkpatrick, H. A., Goeden, M. A., Rose, D. J., Mau, B., and Shao, Y. (1997). The complete genome sequence of *Escherichia coli* K-1. *Science*, 277, 1453-62.

[23] Lewis, N. E., Liu, X., Li, Y., Nagarajan, H., Yerganian, G., O'Brien, E., Bordbar, A., Roth, A. M., Rosenbloom, J., Bian, C., Xie, M., Chen, W., Li, N., Baycin-Hizal, D., Latif, H., Forster, J., Betenbaugh, M. J., Famili, I., Xu, X., Wang, J., and Palsson, B. O. (2013). Genomic landscapes of Chinese hamster ovary cell lines as revealed by the *Cricetulus griseus* draft genome. *Nat Biotechnol*, 31, 759-65.

[24] Singh, S. K., Mishra, A., Yadav, D., Budholiya, N., and Rathore, A. S. (2020). Understanding the mechanism of copurification of "difficult to remove" host cell proteins in rituximab biosimilar products. *Biotechnol Prog*, 36, e2936.

[25] Dovgan, T., Golghalyani, V., Zurlo, F., Hatton, D., Lindo, V., Turner, R., Harris, C., and Cui, T. (2021). Targeted CHO cell engineering approaches can reduce HCP-related enzymatic degradation and improve mAb product quality. *Biotechnol Bioeng*, 118, 3821-31.

[26] Gao, S. X., Zhang, Y., Stansberry-Perkins, K., Buko, A., Bai, S., Nguyen, V., and Brader, M. L. (2011). Fragmentation of a highly purified monoclonal antibody attributed to residual CHO cell protease activity. *Biotechnol Bioeng*, 108, 977-82.

[27] Quarmby, V., Phung, Q. T., and Lill, J. R. (2018). MAPPs for the identification of immunogenic hotspots of biotherapeutics; an overview of the technology and its application to the biopharmaceutical arena. *Expert Rev Proteomics*, 15, 733-48.

[28] Wang, F., Li, X., Swanson, M., Guetschow, E., Winston, M., Smith, J. P., Hoyt, E., Liu, Z., Richardson, D., Bu, X., Jawa, V., and Variankaval, N. (2022). Holistic analytical characterization and risk assessment of residual host cell protein impurities in an active pharmaceutical ingredient synthesized by biocatalysts. *Biotechnol Bioeng*, 119, 2088-104.

[29] Verthelyi, D., and Wang, V. (2010). Trace levels of innate immune response modulating impurities (IIRMIs) synergize to break tolerance to therapeutic proteins. *PLoS One*, 5, e15252.

[30] Haile, L. A., Puig, M., Kelley-Baker, L., and Verthelyi, D. (2015). Detection of innate immune response modulating impurities in therapeutic proteins. *PLoS One*, 10, e0125078.

[31] Chamberlain, P., and Rup, B. (2020). Immunogenicity risk assessment for an engineered human cytokine analogue expressed in different cell substrates. *AAPS J*, 22, 65.

[32] Swiech, K., Picanco-Castro, V., and Covas, D. T. (2017). Production of recombinant coagulation factors: Are humans the best host cells? *Bioengineered*, 8, 462-70.

[33] Levy, N. E., Valente, K. N., Lee, K. H., and Lenhoff, A. M. (2016). Host cell protein impurities in chromatographic polishing steps for monoclonal antibody purification. *Biotechnol Bioeng*, 113, 1260-72.

[34] Aoyama, S., Matsumoto, Y., Mori, C., and Sota, K. (2022). Application of novel mixed mode chromatography (MMC) resins having a hydrophobic modified polyallylamine ligand for monoclonal antibody purification. *J Chromatogr B*, 1191, 123072.

[35] Champion, K., Madden, H., Dougherty, J., and Shacter, E. (2005). Defining your product profile and maintaining control over it. *Bioprocess Int.*, Sept, 52-7.

[36] Jones, A. J. S. Sensitive detection and quantitation of protein contaminants. In: rDNA Products. In: Phillips M, editor. The impact of chemistry on biotechnology. 1990/01/01 ed: *Am Chem Soc*, p 193–201.

[37] Eaton, L. C. (1995). Host cell contaminant protein assay development for recombinant biopharmaceuticals. *J Chromatogr A*, 705, 105-14.

[38] Zhang, Q., Goetze, A. M., Cui, H., Wylie, J., Trimble, S., Hewig, A., and Flynn, G. C. (2014). Comprehensive tracking of host cell proteins during monoclonal antibody purifications using mass spectrometry. *Mabs*, 6, 659-70.

[39] Chon, J. H., and Zarbis-Papastoitsis, G. (2011). Advances in the production and downstream processing of antibodies. *Nat Biotechnol*, 28, 458-63.

[40] Bracewell, D. G., Francis, R., and Smales, C. M. (2015). The future of host cell protein (HCP) identification during process development and manufacturing linked to a risk-based management for their control. *Biotechnol Bioeng*, 112, 1727-37.

[41] Hogwood, C. E., Bracewell, D. G., and Smales, C. M. (2014). Measurement and control of host cell proteins (HCPs) in CHO cell bioprocesses. *Curr Opin Biotechnol*, 30, 153-60.

[42] Berkelman, T., Harbers, A., and Bandhakavi, S. (2015). 2-D Western blotting for evaluation of antibodies developed for detection of host cell protein. *Methods Mol Biol*, 1295, 393-414.

[43] Valente, K. N., Schaefer, A. K., Kempton, H. R., Lenhoff, A. M., and Lee, K. H. (2014). Recovery of Chinese hamster ovary host cell proteins for proteomic analysis. *Biotechnol J*, 9, 87-99.

[44] Jin, M., Szapiel, N., Zhang, J., Hickey, J., and Ghose, S. (2010). Profiling of host cell proteins by two-dimensional difference gel electrophoresis (2D-DIGE): Implications for downstream process development. *Biotechnol Bioeng*, 105, 306-16.

[45] Tscheliessnig, A. L., Konrath, J., Bates, R., and Jungbauer, A. (2013). Host cell protein analysis in therapeutic protein bioprocessing - methods and applications. *Biotechnol J*, 8, 655-70.

[46] Pilely, K., Nielsen, S. B., Draborg, A., Henriksen, M. L., Hansen, S. W. K., Skriver, L., Mørtz, E., and Lund, R. R. (2020). A novel approach to evaluate ELISA antibody coverage of host cell proteins-combining ELISA-based immunocapture and mass spectrometry. *Biotechnol Prog*, 36, e2983.

[47] Yang, F., Walker, D. E., Schoenfelder, J., Carver, J., Zhang, A., Li, D., Harris, R., Stults, J. T., Yu, X. C., and Michels, D. A. (2018). A 2D LC-MS/MS strategy for reliable detection of 10-ppm level residual host cell proteins in therapeutic antibodies. *Anal Chem*, 90, 13365-72.

[48] Zhou, M., Lucas, D. A., Chan, K. C., Issaq, H. J., Petricoin, E. F., 3rd., Liotta, L. A., Veenstra, T. D., and Conrads, T. P. (2004). An investigation into the human serum "interactome". *Electrophoresis*, 25, 1289-98.

[49] Soderquist, R. G., Trumbo, M., Hart, R. A., Zhang, Q., and Flynn, G. C. (2015). Development of advanced host cell protein enrichment and detection strategies to enable process relevant spike challenge studies. *Biotechnol Prog*, 31, 983-9.

[50] Johnson, R. O., Greer, T., Cejkov, M., Zheng, X., and Li, N. (2020). Combination of FAIMS, Protein A depletion, and native digest conditions enables deep proteomic profiling of host cell proteins in monoclonal antibodies. *Anal Chem*, 92, 10478-84.

[51] Chen, I. H., Xiao, H., Daly, T., and Li, N. (2020). Improved host cell protein analysis in monoclonal antibody products through molecular weight cutoff enrichment. *Anal Chem*, 92, 3751-7.

[52] Zhao, B., Abdubek, P., Zhang, S., Xiao, H., and Li, N. (2022). Analysis of host cell proteins in monoclonal antibody therapeutics through size exclusion chromatography. *Pharm Res*, 39, 3029-37.

[53] Wang, Q., Slaney, T. R., Wu, W., Ludwig, R., Tao, L., and Leone, A. (2020). Enhancing host-cell protein detection in protein therapeutics using HILIC enrichment and proteomic analysis. *Anal Chem*, 92, 10327-35.

[54] Pereira Bresolin, I. R. A., Lingg, N., Bresolin, I. T. L., and Jungbauer, A. (2020). Hydrophobic interaction chromatography as polishing step enables obtaining ultra-pure recombinant antibodies. *J Biotechnol*, 324S, 100020.

[55] Nie, S., Greer, T., O'Brien, Johnson, R., Zheng, X., Torri, A., and Li, N. (2021). Simple and sensitive method for deep profiling of host cell proteins in therapeutic antibodies by combining ultra-low trypsin concentration digestion, long chromatographic gradients, and boxcar mass spectrometry acquisition. *Anal Chem*, 93, 4383-90.

[56] Bachi, A., Simo, C., Restuccia, U., Guerrier, L., Fortis, F., Boschetti, E., Masseroli, M., and Righetti, P. G. (2008). Performance of combinatorial peptide libraries in capturing the low-abundance proteome of red blood cells. 2. Behavior of resins containing individual amino acids. *Anal Chem*, 80, 3557-65.

[57] Simó, C., Bachi, A., Cattaneo, A., Guerrier, L., Fortis, F., Boschetti, E., Podtelejnikov, A., and Righetti, P. G. (2008). Performance of combinatorial peptide libraries in capturing the low-abundance proteome of red blood cells. 1. Behavior of mono- to hexapeptides. *Anal Chem*, 80, 3547-56.

[58] Righetti, P. G., Candiano, G., Citterio, A., and Boschetti, E. (2015). Combinatorial peptide ligand libraries as a "Trojan Horse" in deep discovery proteomics. *Anal Chem*, 87, 293-305.

[59] Thulasiraman, V., Lin, S., Gheorghiu, L., Lathrop, J., Lomas, L., Hammond, D., and Boschetti., E. (2005). Reduction of the concentration difference of proteins in biological liquids using a library of combinatorial ligands. *Electrophoresis*, 26, 3561-71.

[60] Huhn, C., Ruhaak, L. R., Wuhrer, M., and Deelder, A. M. (2011). Hexapeptide library as a universal tool for sample preparation in protein glycosylation analysis. *J Proteomics*, 75, 1515-28.

[61] Rivers, J., Hughes, C., McKenna, T., Woolerton, Y., Vissers., P. C., Langridge, J. I., and Beynon, R. J. (2011). Asymmetric proteome equalization of the skeletal muscle proteome using a combinatorial hexapeptide library. *PLoS ONE*, 6, e28902.

[62] Li, L. (2015). Dynamic range compression with ProteoMiner: principles and examples. *Meth Mol Biol*, 1295, 99-108.

[63] Brgles, M., Clifton, J., Walsh, R., Huang, F., Rucevic, M., Cao, L., Hixon, D., Muller, E., and Jovic, D. (2011). Selectivity of monolithic supports under overloading conditions and their use for separation of human plasma and isolation of low abundance proteins. *J Chromatogr*, 1218, 2389-95.

[64] Masuda, Y., Tsuda, M., Hashikawa-Muto, C., Takahashi, Y., Nonaka, K., and Wakamatsu, K. (2019). Cation exchange chromatography performed in overloaded mode is effective in removing viruses during the manufacturing of monoclonal antibodies. *Biotechnol Prog*, 35, e2858.

[65] Muca, R., Marek, W., Zurawski, M., Piatkowski, W., and Antos, D. (2017). Effect of mass overloading on binding and elution of unstable proteins in hydrophobic interaction chromatography. *J Chromatogr*, 1492, 79–88.

[66] Cho, W. (2015). Displacement phenomena in lectin affinity chromatography. *Anal Chem*, 8, 9612-20.

[67] Evans, S., Holstein, M., and Cramer, S. M. (2011). Detection of trace proteins in multicomponent mixtures using displacement chromatography. *Anal Chem*, 83, 4184-92.

[68] Lee, P. Y., Saraygord-Afshari, N., and Low, T. Y. (2020). The evolution of two-dimensional gel electrophoresis - from proteomics to emerging alternative applications. *J Chromatogr A*, 1615, 460763.

[69] Li, X., Wang, W., and Chen, J. (2017). Recent progress in mass spectrometry proteomics for biomedical research. *Sci China Life Sci*, 60, 1093-1113.

[70] Rotello, R. J., and Veenstra, T. D. (2021). Mass spectrometry techniques: Principles and practices for quantitative proteomics. *Curr Protein Pept Sci*, 22, 121-33.

[71] Candiano, G., Dimuccio, V., Bruschi, M., Santucci, L., Gusmano, R., Boschetti, E., Righetti, P. G., and Ghiggeri, G. M. (2009). Combinatorial peptide ligand libraries for urine proteome analysis : investigation of different elution systems. *Electrophoresis*, 30, 2405-11.

[72] Ernoult, E., Bourreau, A., Gamelin, E., and Guette, C. J. (2010). A proteomic approach for plasma biomarker discovery with iTRAQ labelling and OffGel fractionation. *Biomed Biotechnol*, 2010, 927917.

[73] Righetti, P. G., and Boschetti, E. (2015). Sample treatment methods involving combinatorial peptide ligand libraries for improved proteomes analyses. *Meth Mol Biol*, 1243, 55-82.

[74] Peffers, M. J., McDermott, B., Clegg, P. D., and Riggs, C. M. (2015). Comprehensive protein profiling of synovial fluid in osteoarthritis following protein equalization. *Osteoarthritis Cartilage*, 23, 1204-13.

[75] Anderson, J. R., Phelan, M. M., Rubio-Martinez, L. M., Fitzgerald, M. M., Jones, S. W., Clegg, P. D., and Peffers, M. J. (2020). Optimization of synovial fluid collection and processing for NMR metabolomics and LC-MS/MS proteomics. *J Proteome Res*, 19, 2585-97.

[76] Lavoie, R. A., di Fazio, A. R., Blackburn, K., Goshe, M. B., Carbonell, R. G., and Menegatti, S. (2019). Targeted Capture of Chinese Hamster Ovary Host Cell Proteins: Peptide Ligand Discovery. *Int J Mol Sci*, 20, 1729-40.

[77] Mörtstedt, H., Makower, A., Edlund, P. O., Sjöberg, K., and Tjernberg, A. (2020). Improved identification of host cell proteins in a protein biopharmaceutical by LC-MS/MS using the ProteoMiner enrichment kit. *J Pharm Biomed Anal*, 185, 113256.

[78] Yen, E. S., Hu, Y., Molden, R., Qiu, H., and Li, N. (2023). Identification and quantification of a problematic host cell protein to support therapeutic protein development. *J Pharm Sci*, 112, 673-9.

[79] Molden, R., Hu, M., Yen, S., Saggese, D., Reilly, J., Mattila, J., Qiu, H., Chen, G., Bak, A., and Li, N. (2021). Host cell protein profiling of commercial therapeutic protein drugs as a benchmark for monoclonal antibody-based therapeutic protein development. *mAbs*, 13, 1955811.

[80] Chen, I. H., Xiao, H., and Li, N. (2020). Improved host cell protein analysis in monoclonal antibody products through ProteoMiner. *Anal Biochem*, 610, 113972.

[81] Zhang, S., Xiao, H., Molden, R., Qiu, H., and Li, N. (2020). Rapid polysorbate 80 degradation by liver carboxylesterase in a monoclonal antibody formulated drug substance at early stage development. *J Pharmaceut Sci*, 109, 3300-07.

[82] Zhang, S., Riccardi, C., Kamen, D., Reilly, J., Mattila, J., Bak, H., Xiao, H., and Li, N. (2022). Identification of the specific causes of polysorbate 20 degradation in monoclonal antibody formulations containing multiple lipases. *Pharm Res*, 39, 75-87.

[83] Li, X., Wang, F., Li, H., Richardson, D. D., and Roush, D. J. (2022). The measurement and control of high-risk host cell proteins for polysorbate degradation in biologics formulation. *Antib Ther*, 5, 42-54.

[84] Bee, J. S., Tie, L., Johnson, D., Dimitrova, M. N., Jusino, K. C., and Afdahl, C. D. (2015). Trace levels of the CHO host cell protease cathepsin D caused particle formation in a monoclonal antibody product. *Biotechnol Prog*, 31, 1360-9.

[85] Lewis, N. E., Liu, X., Li, Y., Nagarajan, H., Yerganian, G., O'Brien, E., Bordbar, A., and Roth, A. M. (2021). Targeted CHO cell engineering approaches can reduce HCP-related enzymatic degradation and improve mAb product quality. *Biotechnol Bioeng*, 118, 3821-31.

[86] Gibofsky, A., Evans, C., and Strand, V. (2022). Provider and patient knowledge gaps on biosimilars: insights from surveys. *Am J Manag Care*, 28, 227-33.

[87] Madsen, J. A., Farutin, V., Carbeau, T., Wudyka, S., Yin, Y., Smith, S., Anderson, J., and Capila, I. (2015). Toward the complete characterization of host cell proteins in biotherapeutics via affinity depletions, LC-MS/MS, and multivariate analysis. *Mabs*, 7, 1128-37.

Chapter 5

The Function of Cyclo (L-Phenylalanine-L-Proline) on the Proteome of *Staphylococcus Aureus*

Duiyuan Ai[1,*], PhD
Weibing Zhang[1], PhD
and Jianmin Yun[1], PhD
College of Food Science and Engineering, Gansu Agricultural University, Lanzhou, Gansu, People's Republic of China

Abstract

Cyclo (L-phenylalanine-L-proline) (cFP) plays an extracellular quorum sensing (QS) signal molecule role that coordinates communication in Gram-negative bacteria. However, the effect of cFP on Gram-positive bacteria remains unknown. Therefore an isobaric tags for relative and absolute quantitation (iTRAQ) proteomic experiment was designed to elucidate whether cFP influences protein expression in Staphylococcus aureus (S. aureus). The proteomics method was used to detect untreated (control) and S. aureus treated with cFP for 12 h. Samples were then processed by liquid phase tandem mass spectrometry (LC-MS/MS) and analysed using iTRAQ. The results identified 1296 proteins from the S. aureus CGMCC 1.1861 proteome. Twenty-two proteins, including some two-component regulatory systems (TCRS), were associated with signal transduction. Differential expression analysis revealed that only 43 proteins were up-regulated and 41 proteins were down-regulated by cFP. The most significantly different pathways were amino acid metabolism, fatty acid degradation, and metabolism of cofactors and vitamins. Results

* Corresponding Author's Email: aidy@gsau.edu.cn.

In: Advances in Biology. Volume 5
Editor: Charles D. Grant
ISBN: 979-8-89113-441-6

showed that cFP down-regulated virulence factors, up-regulated lipid and amino acid metabolism, promoted acetylation and phosphorylation, and decreased alcohol dehydrogenase expression. A total of 12 significantly differentially expressed proteins (DEPs) were related to signal transduction. Among them, Rot (Q9RFJ6) and SarR (Q9F0R1), which can inhibit transcription of the Agr system, were up-regulated, whereas virulence factors such as ESAT-6 protein A (Q2G189), phenol soluble modulin (Psm, Q2FZA4), and a peptide ABC transporter permease (Q2G168) were down-regulated. AgrA (Q2FWM4) was down-regulated by cFP in S. aureus. All these phenomenon indicate that cFP reduced AgrA and the expression of some exotoxins but increased Rot and SarR expression.

Keywords: cyclic dipeptide, quorumsensing, two-component regulatory systems, signal pathway, pathogenicity

Introduction

Quorum sensing (QS) is a crucial social behavior of bacteria according to their population densities (Banerjee and Arun 2017). QS Involves the production of and response to extracellular signaling molecules called autoinducers (AIs) (Kumari et al., 2006). Some common QS phenotypes include group behaviors such as competence, colonization, motility, biofilm formation, virulence factor expression, and stress responses (Kavanaugh and Horswill 2016; Eickhoff et al., 2018; Bettenworth et al., 2019; Jiang et al., 2019; Zhao et al., 2019). Certain bacteria are also able to use QS to regulate bioluminescence, nitrogen fixation, and sporulation (Kavanaugh and Horswill 2009; Rutherford and Bassler 2012).

Some genes are modified in response to a threshold AIs concentration that depends on high cell density (Daniels et al., 2006). In Gram-negative bacteria, typical QS molecules are N-acyl homoserine lactones (AHL) or other molecules whose production depends on S-adenosylmethionine as a substrate(Wei et al., 2011). In general, AHLs can bind directly to transcription factors (TFs) to regulate gene expression (Bassler 1999). Gram-positive bacteria typically use two-component regulatory systems (TCRS) with autoinducing peptides (AIP) such as oligopeptides for QS (Rutherford and Bassler 2012). AIP binds to a membrane-bound histidine kinase receptor (HK) to auto-phosphorylate and activate their cognate cytoplasmic response regulator. The phosphorylated response regulator activates transcription of genes in the QS regulon. Another possible mechanism is direct AIP binding to a TF that initiates or inhibits transcription (Rutherford and Bassler 2012).

In Gram-positive QS circuits, the pro-AIP, transporter, HK, and TF are typically encoded in an operon (Ji et al., 1995; Peterson et al., 2000). Expression of this operon is activated by the phosphorylated response regulator, initiating an auto-inducing feed-forward loop that synchronizes the QS response. The Staphylococcus aureus accessory gene regulator (Agr) system is a model Gram-positive QS system also known as a TCRS signal transduction pathway (Thoendel et al., 2011). When the concentration of the signal molecule reaches a threshold, the membrane-bound HK, AgrC is phosphorylated, stimulating the Agr system. The AIP precursor is synthesized via AgrD, processed into the mature AIP octapeptide, and transported out of the cell by AgrB. The AIP is then detected by AgrC and the TF, AgrA is phosphorylated by phosphorylated AgrC. AgrA can then activate the agr P2 (RNAII) and P3 (RNAIII) promoters. RNAIII post-transcriptionally activates expression of virulence factors and represses Rot, which is the main repressor of toxin genes encoding virulence factors. Gram-positive bacteria can initiate pathogenic processes via QS circuits that allow for detection of cell population density. The final consequence of this QS regulatory cascade is down-regulation of surface virulence factors such as protein A (spa) and up-regulation of secreted virulence factors such as α-toxin and σ-toxin (Rutherford and Bassler 2012; Kavanaugh and Horswill 2016). In S. aureus, some researchers have suggested that the Agr system inhibits biofilm formation (Vuong et al., 2000; Boles and Horswill 2008). This leads bacteria to establish biofilm communities at low cell densities but terminate biofilm production and decrease surface proteins and adhesions once they reach high cell densities. In this way, the cell is able to secrete virulence factors that facilitate its dispersal and invasion (Yarwood et al., 2004; Boles and Horswill 2008).

Cyclo (L-phenylalanine-L-proline) (cFP) is a cyclic dipeptide formed by the condensation of two amino acids also called 2, 5-diketopiperazines. It consists of a stable, six-membered ring structure, which includes one hydrogen donor and four acceptor bonds (Hu et al., 2019). This makes it an active ligand with key roles in protein binding. It is a secondary metabolite produced by numerous fungi as well as Gram-positive and Gram-negative bacteria (Prasad 1995; Huang et al., 2010; Brack et al., 2014; Mishra et al., 2017, Mishra et al., 2019). In Gram-negative bacteria such as Pseudomonas spp., cFP was shown to be part of a novel class of AIs that uses lux-based AHL biosensors (Bellezza et al., 2014). Furthermore, cFP can reduce bioluminescence and biofilm formation (Kim et al., 2013); induce the virulence factor OmpU to influence host–pathogen interactions in Vibrio vulnificus (Park et al., 2006); up-regulate pathogenicity and toxin proteins such as Ctx in Vibrio cholera (Bina et al., 2013); decrease toxic shock syndrome toxin-1 (TSST-1) expression (Li et al., 2011); and inhibit serine/threonine protein kinase (Akt) (Hong et al., 2008). cFP is a very small molecule that can be easily obtained by fermentation or synthesis (Park et al.,2006). Its multiple uses include clinical applications as an antibiotic prophylaxis and inhibition of biofilm formation and multi-drug resistance in bacteria (Rhee et al.,2004; Kimet al.,2015). Recently it was found to be toxic to tumor cells and can inhibit virus and human innate immunity (Braunset al., 2004; Braunset al., 2005; Chen et al., 2018; Kwaket al., 2018; Lee et al., 2018). However, the study of cFP has focused mainly on LuxR-mediated QS systems in bacteria, while its signal transduction mechanism, binding receptors, and in particular its effect on Gram-positive bacteria remain unknown. One transcriptomic study of V. vulnificus indicated that cFP up-regulated transport and metabolism of inorganic molecules and down-regulated glycolysis, anaerobic energy metabolism, and the pyruvate oxidase system (Kim et al., 2013). It was also found that cFP acts as a quorum quenching agent to inhibit Agr (Li et al., 2011; Tang and Zhang 2014).

Isobaric tags for relative and absolute quantitation (iTRAQ) is a proteome quantification technique that can be used to analyze up to eight samples in one experiment (Wiese et al., 2007) and can be used for high throughput generation of abundant, repeatable, high resolution data with high quantitative accuracy (Chai and Zhao 2017). This technology, developed by Applied Biosystems Inc., has been widely used since it was first proposed by Rosset al., (2004) at a mass spectrometry (MS) conference. In 2009, more than 150 research articles reported using this method, and it is now the most common procedure for proteomic analysis. iTRAQ labeling coupled with multidimensional liquid chromatography and MS analysis (LC-MS/MS) is a gel-free quantitative proteomics technology that uses amine-specific isobaric tags to compare the intensity of reporter ions of labeled peptides and infer quantitative values for corresponding proteins (DeSouza et al., 2005, Zhang et al., 2017). In this study, iTRAQ was conducted to elucidate the influence of cFP on protein expression in S. aureus. This work provides crucial clues about how cFP affects the QS mechanismvia TCRS, especially the main QS system AgrC/A, through analysis of differential protein expression.

Materials and Methods

Bacteria and Reagents

*Staphylococcus aureus*subsp.*Aureus*CGMCC1.1861 (ATCC 6538P; *S. aureus* 1.1861), supplied by the China General Microbiological Culture Collection Center (CGMCC, Beijing, China), is a positive control strain used in medicine and microbial hygiene tests(Holowachuk et al., 2003; Holowachuk et al., 2004).cFP(IUPAC name:(3S,8aS)-3-benzyl-2,3,6,7,8,8a-hexahyd ropyrrolo[1,2-a]pyrazine-1,4-dione,CAS number: 3705-26-8) was synthesized by Nanjing Peptide Biotech Ltd. (Nanjing, China). *S. aureus* 1.1861was previously shown to be sensitive to cFP in growth curves (Duiyuan2017). It was grown in basic nutrient beef extract peptone media in Erlenmeyer flasks at 37 °C with shaking for 18 h. After continuous culture for two generations, cells were diluted to an optical density at 600 nm (OD_{600}) of 0.1 per mL and 100 mL was separated into one Erlenmeyer flask for each condition. The prepared samples were divided into two groups. Control (CK) samples of *S. aureus* treated with 0.1 mL dimethyl sulfoxide (DMSO) were cultured at 37 °C withshaking for 12 min basic nutrient beef extract peptone media. Treatment group (cFP) samples of *S. aureus* treated with cFP dissolved in 0.1 mL DMSO at a final concentration of 10 mM were cultured for 12 hin the same conditions. Each group included three independent replicates (i.e., CK-1, CK-2, and CK-3 in the control group and cFP-1, cFP-2, and cFP-3 in the treatment group). Both groups were collected by centrifugation at 1200 x *g* for 5 min at 4 °C and frozen in liquid nitrogen until further analysis.

iTRAQ Proteomic Experiment

The iTRAQ workflow is illustrated in Figure 1. Total protein was extracted from processed samples using the cold acetone method (Xiao et al., 2017). Samples were ground to a powder in liquid nitrogen, dissolved in 2 mL lysis buffer (8 M urea, 2% SDS, and 1x protease inhibitor cocktail (Roche Ltd., Basel, Switzerland)), sonicated on ice for 30 min and centrifuged at 13,000 rpm for 30 min at 4 °C. The supernatant was transferred to a fresh tube. For each sample, proteins were precipitated with ice-cold acetone at -20 °C overnight. The precipitates were washed with acetone three times and re-dissolved in 8 M urea with sonication on ice. Protein quality was examined by SDS-PAGE.

A Bradford protein concentration assay (BCA) kit (P0010S, Beyotime Institute of Biotechnology, Beijing, China) was used to determine the protein concentration of the supernatant. A total of 100 μg protein was transferred to a new tube and adjusted to a final volume of 100 μL with 8 M urea. A total of 11 μL of 1 M DL-dithiothreitol was added and samples were incubated at 37 °C for 1 h. Then, 120 μL of 55 mM iodoacetamide was added and the sample was incubated for 20 min at room temperature and protected from light. For each sample, proteins were precipitated with ice-cold acetone and re-dissolved in 100 μL triethylammonium bicarbonate buffer. Proteins were then tryptic digested with sequence-grade modified trypsin (Promega, Madison, WI, USA) at 37 °C overnight. The resulting peptide mixture was covalently labeled with stable isotope molecules with the following tags introduced from iTRAQ-specific reagents to the N-terminus and side chain amines of peptides according to the manufacturer's instructions: control group CK-1 with a 113 mass isotope tag; CK-2 with a 114 tag; CK-3 with a 115 tag; and treatment group cFP-1 with a 116 tag; cFP-2 with a 117 tag; and cFP-3 with a 118 tag. The labeled samples were combined and dried in a vacuum (Zhang et al., 2017).

Labeled samples were combined and subjected to strong cation exchange fractionation on a column connected to a high-performance liquid chromatography system. The peptide mixture was dissolved in buffer A (20 mM ammonium formate in water, pH adjusted to 10.0 with ammonium hydroxide) and fractionated by high pH separation using an Ultimate 3000 system (Thermo Fisher Scientific, Waltham, MA, USA) connected to a reverse-phase column. Peptide fractions were resuspended in 30 μl solvent C (0.1% formic acid in water) or D (0.1% formic acid in acetonitrile), separated by nanoLC, and analyzed by on-line electrospray tandem mass spectrometry. Experiments were performed on an Easy-nLC 1000 system (Thermo Fisher Scientific) connected to an Orbitrap Fusion Tribrid mass spectrometer (Thermo Fisher Scientific) equipped with an online nano-electrospray ion source. Survey full-scan mass spectra (m/z 350–1550) were acquired with a mass resolution of 120 K, followed by sequential high energy collisional dissociation tandem mass spectrometry scans with a resolution of 30 K. The isolation window was set to 1.6 Da. The AGC target was set to 400,000. The fixed first mass was set to 110. In all cases, one microscan was recorded using a dynamic exclusion of 45 s (Chai and Zhao 2017).

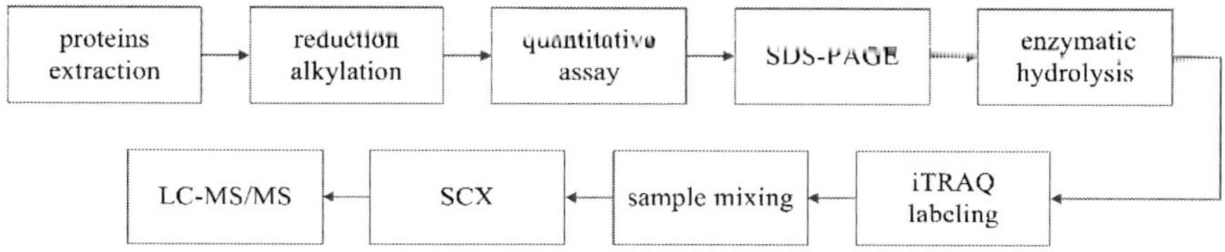

Figure 1. iTRAQ workflow. Abbreviations: SDS-PAGE, sodium dodecyl sulfate polyacrylamide gel electrophoresis; LC-MS/MS, liquid phase tandem mass spectrometry analysis; SCX, strong cation exchange chromatography separation.

Database Analysis

Mass spectrometry data were transformed into Mascot generic format (MGF) files using Proteome Discovery 1.2 (Thermo, Pittsburgh, PA, USA) and analyzed using the Mascot search engine version 2.3.2 (Matrix Science, London, UK) (Figure 2). The Mascot database was set up for protein identification using the reference transcriptome or NCBInr/SwissProt/Uniprot/IPI database. Mascot was searched with a fragment ion mass tolerance of 0.050 Da and a parent ion tolerance of 10.0 ppm. NCTC 8325 is the only reference proteome in Uniprot (UP000008816, strain NCTC 8325, protein count: 2889, https://www.uniprot.org/proteomes/UP000008816) and is well studied and annotated (Gillaspy et al., 2006). Therefore, it was selected as the main reference proteome for our data search results. The Kyoto Encyclopedia of Genes and Genomes (KEGG) (Kanehisa et al., 2008) pathway enrichment analysis was used to identify differentially expressed proteins (DEPs) significantly enriched in metabolic or signal transduction pathways when compared with the entire proteome. The following formula was used to calculate P-values:

$$P = 1 - \sum_{i=0}^{m-1} \frac{\binom{M}{i}\binom{N-M}{n-i}}{\binom{N}{n}}$$

where N is the number of all genes with a KEGG annotation; n is the number of DEPs in N; M is the number of genes annotated to specific pathways; and m is the number of DEPs in M. The calculated P-values were adjusted using the false discovery rate (FDR) correction with FDR ≤ 0.05 set as the threshold. Pathways meeting this condition were defined as pathways significantly enriched in DEPs.

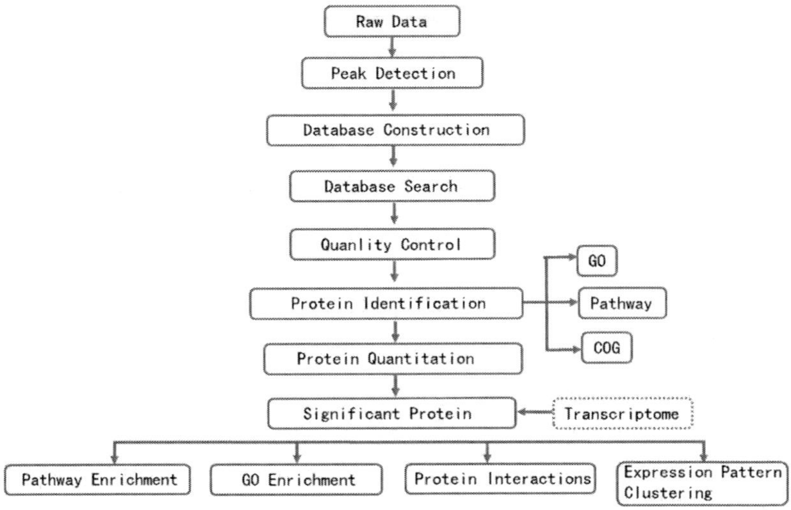

Figure 2. Proteomic analysis workflow. The raw data were transformed and analyzed using Mascot software. These data were then assessed for protein identification using the NCBInr/SwissProt/Uniprot/IPI database. Afterwards, identified proteins were quantified to identify significantly differentially expressed proteins (DEPs). Finally, DEPs were analysed for KEGG pathway enrichment, Gene Ontology (GO) enrichment, and protein interactions to identify mechanisms at the overall proteome level.

Results

Protein Identification and Quantification

Proteomics experiments referring to the NCBInr/SwissProt/Uniprot/IPI database identified 1296 S. aureus proteins in both control and treatment groups. Mascot search results were averaged using medians and quantified. Proteins with a fold change > 1.2 or < 0.83 and an unadjusted significance level of $P < 0.05$ were considered differentially expressed. In total, 1153 proteins were quantitative statistics and annotated by KEGG. Of these, 43 proteins were significantly up-regulated and 41 proteins were down-regulated in cFP treatment groups compared with controls (P< 0.05; Table 1). These results revealed clues about target proteins influenced by cFP. The adenosylmethionine-8-amino-7-oxononanoate aminotransferase protein BioA (Q2FVJ6) was the most up-regulated protein, followed by the biotin synthase protein BioB and four ribonuclease M5 proteins. A total of 12 proteins were directly related to transcription and signal transduction among these 84DEPs, including the up-regulated proteins Rot (Q9RFJ6), acetyl-CoA acetyltransferase (Q2G0K2), SarR (Q9F0R1), MalR (A0A0Y9ARI8), MalR (Q2FY63), Transcription factor (TF) (Q2FWG9), and PhoP (Q2FXN6) and the down-regulated proteins ATP-binding Cassette (ABC, Q2G168), RpiR TF (Q2FVV2), EsxA (Q2G189), TF (Q2G2H0) and phenol soluble modulin beta 1 (Psm, Q2FZA4) (Table 2).

Pathway Annotation and Enrichment

In this study, 522 proteins from the treated and control groups were found to be associated with known pathways. The most enriched pathway was carbon metabolism, which included 77 proteins, followed by biosynthesis of amino acids with 61 proteins, ribosome with 54 proteins, purine with 38 proteins, pyruvate with 37 proteins, pyrimidine with 34 proteins, and glycolysis/ gluconeogenesis with 33 proteins. In addition, 22 proteins were related to signal transduction. Nineteen ABC transporters and 11 proteins related to S. aureus infection were identified.

Furthermore, we identified, compared, and predicted the possible functions of identified proteins and determined their functional classification using the Clusters of Orthologous Groups of proteins (COG) database (Figure 3.). In total, 1038 proteins were annotated and categorized into 20 groups (C–V). Abundant clusters included J: Translation, ribosomal structure and biogenesis with 123 proteins; C: Energy production and conversion with 87 proteins; E: Amino acid transport and metabolism with 87 proteins; K: Transcription with 86 proteins; and Carbohydrate transport and metabolism with 78 proteins. A total of 29 proteins were associated with signal transduction mechanisms.

Table 1. Significantly differentially expressed proteins between control and cFP-treated S. aureus

GeneID	Control	cFP	log2_FC (cFP/C)	P-value	FDR	Description
Q2FVJ6	0.9643	2.5560	1.4063	0.0003	0.0579	adenosylmethionine--8-amino-7-oxononanoate aminotransferase BioA
Q2FVJ7	0.8913	1.8340	1.0410	0.0005	0.0701	biotin synthase BioB
A0A0H2I6U0	1.0313	1.9283	0.9028	0.0256	0.2943	ribonuclease M5
G5JJ45	1.0313	1.9283	0.9028	0.0256	0.2943	ribonuclease M5
A0A1E5JK13	1.0313	1.9283	0.9028	0.0256	0.2943	ribonuclease M5
Q2G0T1	1.0313	1.9283	0.9028	0.0256	0.2943	ribonuclease M5
Q9RFJ6	0.9387	1.6023	0.7715	0.0001	0.0420	HTH-type transcription factor (TF) Rot
Q2G0K2	0.9647	1.3957	0.5329	0.0013	0.1087	acetyl-CoA acetyltransferase
Q2FZW2	1.2047	1.7410	0.5313	0.0458	0.3567	NifU-like protein, Nitrogen fixation protein NifU
Q2FY46	0.8717	1.2583	0.5297	0.0348	0.3185	exodeoxyribonuclease 7 large subunit
Q2FZU3	1.0110	1.3960	0.4655	0.0038	0.1835	glycerophosphodiesterphosphodiesterase
Q2G1P3	1.1097	1.5273	0.4609	0.0439	0.3501	oleatehydratase
Q2G0V3	0.8693	1.1930	0.4566	0.0118	0.2372	cystathionine beta-lyase
Q2G1U0	0.9663	1.3233	0.4536	0.0090	0.2372	chain a, molecular mechanism and evolution of guanylate kinase regulation by (p)ppgpp
Q2G002	1.0747	1.4667	0.4487	0.0478	0.3627	3-dehydroquinase
Q9F0R1	1.1597	1.5650	0.4325	0.0176	0.2821	HTH-type TF SarR
Q2G024	1.1650	1.5607	0.4218	0.0418	0.3463	ribonuclease R
Q2G228	0.9853	1.3023	0.4024	0.0430	0.3495	purine nucleoside phosphorylase
A0A0Y9ARI8	1.1453	1.5100	0.3988	0.0323	0.3133	Maltose operon transcriptional repressor MalR, LacI family
Q2FY63	1.1453	1.5100	0.3988	0.0323	0.3133	LacI family TF
Q2FXI5	1.0217	1.3190	0.3685	0.0208	0.2879	peptidase M28
Q2G0R1	0.9127	1.1760	0.3657	0.0199	0.2836	hypoxanthine phosphoribosyltransferase
Q2FY16	1.1943	1.5247	0.3523	0.0440	0.3501	deoxyribonuclease IV
Q2G1D0	0.9067	1.1473	0.3396	0.0101	0.2372	acetyl-CoA acetyltransferase
Q2FWG9	1.0047	1.2707	0.3389	0.0474	0.3621	CsoR family TF

GeneID	Control	cFP	log2_FC (cFP/C)	P-value	FDR	Description
Q2FWY9	1.0640	1.3377	0.3302	0.0041	0.1835	glutamyl-tRNA(Gln) amidotransferase
Q2FYZ0	1.0453	1.3063	0.3216	0.0045	0.1835	glutathione peroxidase
Q2FX00	1.0497	1.3063	0.3156	0.0221	0.2879	glutamine amidotransferase
Q2FXG1	1.0327	1.2837	0.3139	0.0329	0.3133	GTP cyclohydrolase II
A0A1E5JEN8	1.0327	1.2837	0.3139	0.0329	0.3133	3,4-dihydroxy-2-butanone-4-phosphate synthase
Q2FWB8	1.0053	1.2487	0.3127	0.0015	0.1138	purine nucleoside phosphorylase, partial
Q2FXN6	0.9707	1.2030	0.3096	0.0056	0.1870	DNA-binding response TF
Q2FVW7	1.0340	1.2800	0.3079	0.0151	0.2698	acyl-CoAdehydrogenase, N-terminal domainprotein
Q2G101	0.9607	1.1887	0.3072	0.0010	0.0897	Phosphoglyceratemutase family protein
Q2FY51	1.0273	1.2697	0.3055	0.0315	0.3133	dihydrolipoyl dehydrogenase
Q2FWG2	0.9450	1.1630	0.2995	0.0300	0.3112	hydroxyethylthiazole kinase
Q2G202	0.9307	1.1403	0.2931	0.0399	0.3423	Hemoglobin-like protein HbO
Q2G0V7	0.9817	1.1963	0.2853	0.0116	0.2372	carboxylesterase
Q2G0Z5	1.0040	1.2163	0.2768	0.0371	0.3336	NADPH-dependent oxidoreductase
A0A077UYP2	0.9973	1.2070	0.2753	0.0050	0.1856	Free methionine-(R)-sulfoxidereductase
Q2FZY5	1.1173	1.3457	0.2683	0.0305	0.3133	cysteine desulfurase
Q2G241	0.9477	1.1390	0.2653	0.0060	0.1920	glutamate--tRNA ligase
Q2FVT6	1.0117	1.2153	0.2646	0.0197	0.2834	imidazolonepropionase
Q2FW23	0.9600	0.7923	-0.2769	0.0171	0.2811	30S ribosomal protein S5
Q2FUX7	1.0367	0.8457	-0.2938	0.0326	0.3133	arginine deiminaseArcA
Q2G2E9	0.9540	0.7770	-0.2961	0.0215	0.2879	TIGR04141 family sporadically distributed protein
Q2FVG3	0.9147	0.7440	-0.2979	0.0292	0.3102	carboxylesterase
Q2FXH8	0.9857	0.7977	-0.3053	0.0001	0.0318	Uncharacterised protein
Q2G261	1.0030	0.8057	-0.3161	0.0256	0.2943	superoxide dismutase
Q2FZA1	0.9723	0.7793	-0.3192	0.0195	0.2834	GNAT family N-acetyltransferase
Q2G0S5	1.0157	0.8090	-0.3282	0.0001	0.0318	stage V sporulation protein G
Q2FXP2	1.0137	0.8057	-0.3313	0.0088	0.2372	Glyceraldehyde 3-phosphate dehydrogenase
Q2FZP9	0.9203	0.7267	-0.3409	0.0447	0.3529	2'-5' RNA ligase

Table 1. (Continued)

GeneID	Control	cFP	log2_FC (cFP/C)	P-value	FDR	Description
P0A0G2	1.0553	0.8320	-0.3430	0.0424	0.3468	50S ribosomal protein L30, partial
Q2FVT2	1.0050	0.7917	-0.3442	0.0016	0.1153	formimidoylglutamase
Q2FVN2	1.1210	0.8780	-0.3525	0.0481	0.3627	Hsp20/alpha crystallin family protein
Q2FVE0	1.0553	0.8263	-0.3529	0.0008	0.0857	Alkyl hydroperoxidereductaseAhpD
A0A0H3JWD7	0.9387	0.7340	-0.3548	0.0339	0.3179	pathogenicity island protein
Q2FWB1	0.9870	0.7653	-0.3670	0.0001	0.0318	oxidoreductase
A0A0H3JPH0	0.9910	0.7683	-0.3672	0.0469	0.3601	UDP-N-acetylglucosamine 4,6-dehydratase
Q2G1K7	0.8443	0.6463	-0.3855	0.0463	0.3583	capsular polysaccharide synthesis enzyme CapB
Q2G168	0.8967	0.6863	-0.3857	0.0195	0.2834	peptide ABC transporter permease
Q7X2S2	1.0727	0.8140	-0.3981	0.0046	0.1835	carbamate kinase 2 ArcC2
A0A1E5JBN2	1.0297	0.7493	-0.4585	0.0412	0.3457	SH3-like domain protein
Q2FYS7	1.0113	0.7357	-0.4591	0.0004	0.0639	iron-sulfur cluster biosynthesis family protein
Q2FYH7	0.9120	0.6453	-0.4990	0.0128	0.2423	chromosome replication protein DnaD
A0A0U1MPU9	0.9673	0.6643	-0.5421	0.0164	0.2793	Putative staphylococcal protein
Q2FV95	1.1630	0.7903	-0.5573	0.0348	0.3185	L-serine dehydratase, iron-sulfur-dependent subunit alpha
Q2G1I8	1.1310	0.7617	-0.5704	0.0057	0.1870	Uncharacterised protein
A0A1Q8DFZ6	1.0253	0.6903	-0.5707	0.0245	0.2943	Uncharacterised protein
Q2G2J7	1.0233	0.6773	-0.5953	0.0123	0.2372	Uncharacterised protein
A0A133QBW6	1.0233	0.6773	-0.5953	0.0123	0.2372	Uncharacterised protein
Q2FVV2	1.0243	0.6513	-0.6532	0.0028	0.1539	MurR/RpiR family TF
A0A1E5JI47	1.0243	0.6513	-0.6532	0.0028	0.1539	helix-turn-helix domain, rpiR family protein
A0A0H3JNW6	0.9730	0.6153	-0.6611	0.0116	0.2372	Uncharacterised protein
Q2G189	1.1733	0.7290	-0.6866	0.0276	0.3068	virulence factor EsxA
Q2FV94	1.2410	0.6990	-0.8281	0.0120	0.2372	L-serine dehydratase, iron-sulfur-dependent subunit beta
Q2G1K9	1.1357	0.6363	-0.8357	0.0027	0.1539	Bifunctional acetaldehyde-CoA/alcoholdehydrogenase
P0C818	1.0510	0.5730	-0.8752	0.0152	0.2698	PsmA4,Phenol-soluble modulin alpha 4 peptide

GeneID	Control	cFP	log2_FC (cFP/C)	P-value	FDR	Description
A0A0Z0SZ51	0.9583	0.4723	-1.0207	0.0186	0.2821	lipase
Q2FYU5	1.0350	0.4783	-1.1135	0.0085	0.2372	30S ribosomal protein S14
Q2FZA4	1.3880	0.6073	-1.1924	0.0106	0.2372	phenol soluble modulinPsm
Q2G2H0	1.1243	0.4717	-1.2532	0.0106	0.2372	TF
Q2G2F5	0.9063	0.3073	-1.5602	0.0348	0.3185	UDP-GlcNAc 2-epimerase

Table 2. Significantly differentially expressed signal transduction proteins between control and cFP-treated S. aureus

GeneID	Control	cFP	log2_FC (cFP/C)	P-value	FDR	Description
Q9RFJ6	0.9387	1.6023	0.7715	0.0001	0.0419	Rot, HTH-type transcription factor (TF)
Q2G0K2	0.9647	1.3957	0.5329	0.0013	0.1087	acetyl-CoAacetyltransferase, Environmental Information
Q9F0R1	1.1597	1.5650	0.4325	0.0176	0.2821	SarR, HTH-type TF,
A0A0Y9ARI8	1.1453	1.5100	0.3988	0.0323	0.3133	MalR, HTH-type Maltose operon TF repressor, LacI family
Q2FY63	1.1453	1.5100	0.3988	0.0323	0.3133	MalR, Maltose operon transcriptional TF, LacI family
Q2FWG9	1.0047	1.2707	0.3389	0.0474	0.3621	TF,cytosolic protein,CsoR family
Q2FXN6	0.9707	1.2030	0.3096	0.0056	0.1870	PhoP, DNA-binding response, phosphate limitation-assimilation,OmpR family
Q2G168	0.8967	0.6863	-0.3857	0.0195	0.2834	peptide ABC transporter permease
Q2FVV2	1.0243	0.6513	-0.6532	0.0028	0.1538	TF,carbohydrate derivative metabolic process, MurR/RpiR family
Q2G189	1.1733	0.7290	-0.6866	0.0276	0.3067	EsxA, virulence factor extracellular protein A
Q2G2H0	1.1243	0.4717	-1.2532	0.0106	0.2371	TF,DUF1811 domain-containing protein
Q2FZA4	1.388	0.6073	-1.1924	0.0106	0.2371	Psm, phenol soluble modulinbeta 1, Pathogenesis

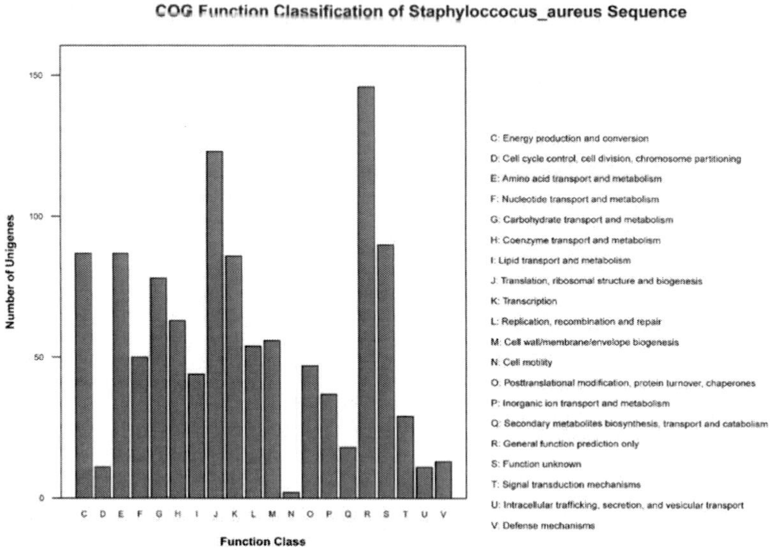

Figure 3. Clusters of Orthologous Groups (COG) of proteins based classification of S. aureus proteins. Using the COG database, 1038 proteins were classified into 20 groups (C–V). Abundant clusters included R: General function with 146 proteins; J: Translation with 123 proteins; S: Function unknown with 90 proteins; C: Energy production and conversion with 87 proteins; and E: Amino acid transport and metabolism with 87 proteins.

Pathway enrichment analysis was based on KEGG pathways and we applied a hypergeometric test to find pathways that were significantly enriched in DEPs between control and treatment groups compared to all identified background proteins. Significantly enriched pathways were used to determine the most important biochemical, metabolic, and signal transduction pathways associated with DEPs (Figure 4.). As shown in Figure 4, based on Q-values, the significantly enriched KEGG class B pathways were amino acid metabolism, lipid metabolism, and metabolism of cofactors and vitamins. Glycine, serine and threonine metabolism ($P = 0.0108$), riboflavin metabolism ($P = 0.0155$), fatty acid degradation ($P = 0.0299$), and valine, leucine and isoleucine degradation ($P=0.0496$) were classified as class C pathways.

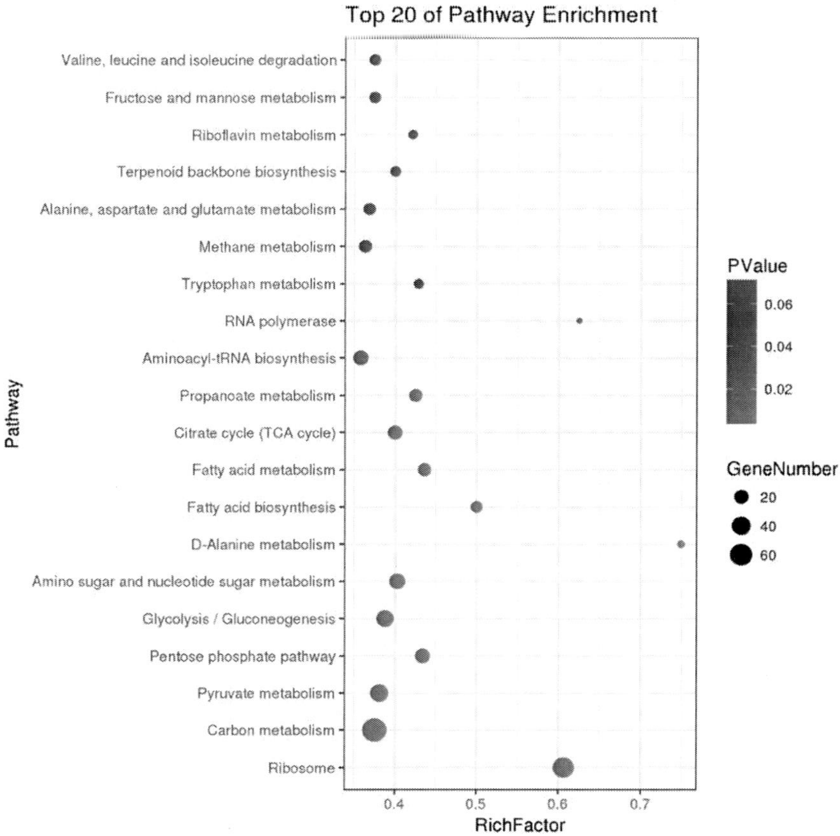

Figure 4. Top 20 enriched pathways of S. aureus proteins influenced by cFP. The Y-axis represents pathways and the X-axis represents their corresponding RichFactor. A larger area indicates a greater number of DEPs, and the color (from white to red; QValue) indicates the greatest difference in protein expression between the control and treatment groups. The most significantly differentially expressed pathways were glycine, serine and threonine metabolism (P = 0.0108), riboflavin metabolism (P = 0.0155), fatty acid degradation (P = 0.0299), and valine, leucine and isoleucine degradation (P = 0.0496).

Discussion

Analysis of cFP treated and control groups revealed that up-regulated proteins included two TCRS receptors, Rot (Q9RFJ6) and SarR (Q9F0R1), and an acetyl-CoA acetyltransferase (Q2G0K2). In addition, some DNA-binding TFs such as MalR (Q2FY63), SAOUHSC_02322 (Q2FWG9), and PhoP (Q2FXN6) were also up-regulated, whereas two TFs (Q2FVV2, Q2G2H0), a peptide ABC (Q2G168), and the virulence factors EsxA (Q2G189) and Psm (Q2FZA4) were down-regulated. Rot, a global TF, negatively regulates the transcription of several known virulence factors, such as lipase (geh), hemolysins (hla and hlb), and proteases (splA–splF, sspB and sspC) (McNamara et al., 2000; Said-Salim et al., 2003), and positively regulates the expression of sarS and other genes, including those encoding cell surface adhesins (clfB, sdrC and spa). Rot and Agr have opposing effects on several genes. Another helix-turn-helix (HTH) type global TF, SarR, negatively regulates sarA transcription (Manna et al., 2001; Manna et al., 2006). Additionally, it negatively regulates expression of RNAII and RNAIII primary transcripts in the Agr locus. EsxA (Q2G189) plays a role as a partial virulence factor belonging to the ESAT-6 secretion system (Sundaramoorthy et al., 2008). Psm (Q2FZA4) is a phenol soluble modulin produced by the Psm-alpha gene cluster that enhances virulence and destruction of white blood cells and increases infectivity (Wang et al., 2007). Psm is promoted by phosphorylated AgrA.

Twenty-two other TCRS proteins were annotated by KEGG pathway analysis, including the histidine kinases NreB (Q2FVM6) and KapB (Q2FZV0), the alkaline phosphatase PhoA (Q2FUY6), and the TF AgrA (Q2FWM4), VraR (Q2FX09), and SrrA (Q2FY79). NreBbelongs to the NreB/NreC family and is involved in the control of dissimilatory nitrate/nitrite reduction in response to low oxygen. PhoA acts in phosphate assimilation in response to phosphate limitation. AgrA is an Agr response protein (Wang et al., 2016). VraR (Q2FX09) functions similarly toa LuxR-type HTH TF by stimulating peptidoglycan synthesis to promote cell wall formation. SrrA (Q2FY79), a global regulator of S. aureus virulence factors in response to environmental oxygen levels, is involved in the TCRS SrrA/SrrB system. SrrA binds to agr, spa, and tst promoters and represses their transcription under low oxygen conditions (Yarwoodet al., 2001; Pragmanet al., 2004). However, in this study SSrA was not differentially expressed between the two groups. In addition, we identified a bacitracin ABC transporter (Q2G0D8) and two acetyl-CoA acetyltransferases (Q2G0K2 and Q2G124) in S. aureus 1.1861.

Agr is the main QS pathway associated with S. aureus virulence and consists of AgrC/A TCRS (Wang et al., 2016). In this study, the HK AgrC was not identified. However, AgrA (Q2FWM4), a crucial regulator in the QS circuit of S. aureus, was identified. Two oligopeptide ABC transporters, OppA (A0A1D4HL93) and OppD (Q2FYQ7), were also found in the S. aureus

1.1861 proteome (Canovas et al., 2017). AIP can bind to AgrC and phosphorylate AgrA to stimulate the expression of RNAIII (Wang et al., 2014). At high cell densities, the Agr system provides positive feed-back to AIP and increases the secretion of virulence factors (Pollitt et al., 2014). Therefore, we hypothesize that, in S. aureus, cFP is transported by membrane transporters such as OppA and OppD andbinds directly to AgrA without the need for AgrC, then activates TF signal transducers and regulators at a large scale. Due to sequence divergence among the four Agr groups (Canovas et al., 2017), it is also possible that another unidentified AgrC receptor recognizescFP in S. aureus 1.1861. In fact, cFP did not increase Agr system positive feed-back but instead significantly down-regulated AgrA (Q2FWM4, cFP/CK log2 fold change = -0.5507). However, cFP did up-regulate Rot (Q9RFJ6; cFP/CK log2 fold change = 0.7715), which is a major repressor of virulence that has opposing effects on certain genes modified by Agr. SarR (Q9F0R1), which competes with AgrA to bind the P2 and P3 promoters and inhibit agrB, agrD, agrC, and agrA transcription, was also up-regulated (cFP/CK log2 fold change = 0.4325). However, EsxA (Q2G189; cFP/CK log2 fold change = -0.6866) and the important virulence factorPsm (Q2FZA4; cFP/CK log2 fold change = -1.1924) that is activated by AgrA, were both down-regulated. These results suggest that cFP did not activate the Agr system but seemed to play the opposite role to that of the octapeptide AIP in the Agr QS system of S. aureus. cFPacted similarly toa competitive inhibitor ora QS quenching agent that binds to AgrC/AgrA to inhibit phosphorylation and suppress cognate accessory geneexpression due to its similar conformation to the octapeptide but not the endogenous AIP. Because four S. aureus subgroups can block QS via their AIPs (AIP1, AIP2, AIP3, and AIP4) (Otto et al., 2001; Tal-Gan et al., 2016), they are cross-inhibitory to the S. aureusAgr system. This indicates that each AgrC can only recognize its endogenous AIP and is repressed by similarly structured molecules. Li et al., (2011) demonstrated a similar phenomenon. These observationssuggest that cFP may interfere with the Agr system, decrease toxin and enzyme expression, increase lipid and amino acid metabolism by promoting acetylation and phosphorylation, and reduce alcohol dehydrogenase expression.

Conclusion

A total of 1296 proteins were identified in the S. aureus 1.1861 proteome. KEGG pathway annotation indicated that the greatest number of proteins (77) were associated with carbon metabolism, whereas 22 proteins belonged to signal transduction, 19 proteins were ABC transporters, and 11 proteins were associated with infectious diseases. Additionally, 1038 genes were annotated by COG, including 29 signal transduction proteins, 11 intracellular trafficking, secretion and vesicular transport proteins, and 13 defense mechanism proteins. In total, 43 proteins were up-regulated and 41 proteins were down-regulated between control and treatment groups. A total of 12 proteins related to transcription and signal transduction were differentially expressed. Among the DEPs, the most significant were those which are known to be involved in amino acid metabolism, lipid metabolism, and metabolism of cofactors and vitamins. The results also showed that AgrA (Q2FWM4) exists in S. aureus 1.1861. Furthermore, cFP down-regulated AgrA and the virulence factors EsxA (Q2G189) andPsm (Q2FZA4) and up-regulated Rot (Q9RFJ6) andSarR (Q9F0R1), which inhibits Agr system transcription. These findings suggest that cFP can reduce Agr system expression.

Disclaimer

None

References

Banerjee G, Arun K. Quorum-sensing network-associated gene regulation in Gram-positive bacteria. *Acta. Microbiol. Immunol. Hung.* (2017) 64:439-453.

Bassler BL. How bacteria talk to each other: regulation of gene expression by quorum sensing. *Curr. Opin. Microbiol.* (1999) 2:582–587. PMID: 10607620.

Bellezza I, Peirce MJ, Minelli A. Cyclic dipeptides: from bugs to brain. *Trends Mol. Med.* (2014) 20:551-558.

Bettenworth V, Steinfeld B, Duin H, Petersen K, Streit W R, Bischofs I, Becker A. Phenotypic Heterogeneity in Bacterial Quorum Sensing Systems. *J. Mol. Biolpii.*: (2019) S0022-2836(19)30244-X. PMID: 31051177.

Bina X R, Taylor D L, Vikram A, Ante V M, Bina J E. Vibrio choleraeToxR down regulates virulence factor production in response to cyclo (Phe-Pro). *MBio*(2013) 4:e00366-13.

Boles BR, Horswill AR. Agr-mediated dispersal of Staphylococcus aureus biofilms. *PloS Pathog.* (2008) 4:e1000052.

Brack C, Mikolasch A, Schauer F. 2, 5-Diketopiperazines Produced by Bacillus pumilus During Bacteriolysis of Arthrobactercitreus. *Mar. Biotechnol.* (2014) 16:385–395.

Brauns S C, Dealtry G, Milne P, Naudé R, Van de Venter M. Caspase -3 activation and induction of PARP cleavage by cyclic dipeptide cyclo(Phe-Pro) in HT-29 cells. *Anticancer Res.* (2005) 25:4197–4202.

Brauns SC, Milne P, Naudé R, Van de Venter M.Selected cyclic dipeptides inhibit cancer cell growth and induce apoptosis in HT-29 colon cancer cells. *Anticancer Res* (2004) 24:1713–1719.

Canovas J, Baldry M, Bojer M S, Andersen P S, Grzeskowiak P K, Stegger M, Damborg P, Olsen CA, Ingmer H. Cross-Talk between Staphylococcus areus and Other Staphylococcal Species via the agr Quorum Sensing System. *Front. Microbiol.* (2016) 7:1733–1745.

Chai Y, Zhao M. ITRAQ-Based Quantitative Proteomic Analysis of the inhibitory effects of polysaccharides from Viscumcoloratum (Kom.) Nakai on HepG2 Cells. *Sci. Rep.* (2017) 7:4596.

Chen C, Ye Y, Wang R, Zhang Y, Wu C, Debnath SC, Ma Z, Wang J, Wu M. Streptomyces nigra sp. nov.is a novel actinobacterium isolated from mangrove soil and exerts a potent antitumor activity in vitro. *Front. Microbiol.* (2018) 9:1587. PMID: 30072967.

Daniels R, Reynaert S, Hoekstra H, Verreth C, Janssens J, Braeken K, Fauvart M, Beullens S, Heusdens C, Lambrichts I, De Vos DE, Vanderleyden J, Vermant J, Michiels J. Quorum signal molecules as biosurfactants affecting swarming in Rhizobium etli. *PNAS USA* (2006) 103:14965–14970.

DeSouza L, Diehl G, Rodrigues M J, Guo J, Romaschin A D, Colgan T J, Siu K W. Search for cancer markers from endometrial tissues using differentially labeled tags iTRAQ and cICAT with multidimensional liquid chromatography and tandem mass spectrometry. *J. Proteome Res.* (2005) 4:377–386.

Duiyuan A. Cyclo-Pro-Phe effects on Staphylococcus aureus growth and energy metabolism. *Res. J. Biotechnol.* (2017) 12:99–104.

Eickhoff MJ, Bassler BL.SnapShot: Bacterial Quorum Sensing. (2018) 174:1328-1328.e1.*PMID*: 30142348.

Gillaspy A F, Worrell V, Orvis J, Roe B A, Dyer D W, Iandolo JJ. The Staphylococcus aureus NCTC 8325 genome. In: Fischetti V (ed) Gram positive pathogens, 2nd edn. Washington DC: *ASM Press* (2006) 381–412.

Holowachuk S A, Bal'a M F. A kinetic microplate method for quantifying the antibacterial properties of biological fluids. *J. Microbiol. Methods* (2003) 55:441–446.

Holowachuk SA1, Bal'a M F, Gerard P D. Nonparallel secretion of antibacterial activity and protein in porcine pancreatic juice. *Pancreas* (2004) 28:E32–38.

Hong S, Moon B H, Yong Y, Shin S Y, Lee Y H, Lim Y. Inhibitory effect against akt by cyclic dipeptides isolated from Bacillus sp. *J. Microbiol. Biotechnol.* (2008) 18:682–685.

Huang R, Zhou X, Xu T, Yang X, Liu Y.Diketopiperazines from marine organisms. *Chem. Biodivers* (2010) 7:2809–2829.

Ji G, Beavis R C. Cell density control of staphylococcal virulence mediated by an octapeptide pheromone. *PNAS USA* (1995) 92:12055–12059.

Jiang Q, Chen J, Yang C, Yin Y, Yao K. Quorum Sensing: A Prospective Therapeutic Target for Bacterial Diseases. *Biomed. Res. Int.* (2019) 2019:2015978. PMID: 3108 0810.

Kanehisa M, Araki M, Goto S, Hattori M, Hirakawa M, Itoh M, Katayama T, Kawashima S, Okuda S, Tokimatsu T, Yamanishi Y. KEGG for linking genomes to life and the environment. *Nucleic Acids Res.* (2008) 36:D480–484.

Kavanaugh J S. Impact of environmental cues on staphylococcal quorum sensing and biofilm development. *J. Biol. Chem.* (2016) 291:12556–12564.

Kim IH, Son JS, Wen Y, Jeong S M, Min G Y, Park N Y, Lee K W, Cho Y J, Chun J, Kim KS. Transcriptomic analysis of genes modulated by cyclo(L-Phenylalanine-L-Proline) in Vibrio vulnificus. *J. Microbiol. Biotechnol.* (2013) 23:1791–1801.

Kim K, Kim N J, Kim S Y, Kim I H, Kim K S, Lee G R. Cyclo(Phe-Pro) produced by the human pathogen Vibrio vulnificus inhibits host innate immune responses through the NF-κB pathway. *Infect. Immun.* (2015) 83:1150–1161.

Kumari A, Pasini P, Deo S K, Flomenhoft D, Shashidhar H, Daunert S. Biosensing systems for the detection of bacterial quorum signaling molecules. *Ana.l Chem.* (2006) 78:7603–7609.

Kwak M K, Liu R. Antimicrobial activity of cyclic dipeptides produced by Lactobacillus plantarum LBP-K10 against multidrug-resistant bacteria, pathogenic fungi, and influenza A virus. *Food Control.* (2018) 85:223–234.

Lee W, Lee SH, Kim M, Moon J S, Kim G W, Jung H G, Kim I H, Oh J E, Jung H E, Lee HK, Ku KB, Ahn DG, Kim SJ, Kim KS, Oh JW. Vibrio vulnificus quorum-sensing molecule cyclo (Phe-Pro) inhibits RIG-I-mediated antiviral innate immunity. *Nat. Commun.* (2018) 9:1606.

Li J, Wang W, Xu SX, Magarvey NA, McCormick JK. Lactobacillus reuteri-produced cyclic dipeptides quench agr-mediated expression of toxic shock syndrome toxin-1 in staphylococci.*Proc. Natl. Acad. Sci. USA* (2011) 108:3360–3365.

Manna A C, Cheung A L. Characterization of sarR, a modulator of sar expression in Staphylococcus aureus. *Infect. Immun.* (2001) 69:885–896.

Manna A C, Cheung A L. Transcriptional regulation of the agr locus and the identification of DNA binding residues of the global regulatory protein SarR in Staphylococcus aureus.*Mol. Microbiol.* (2006) 60:1289–1301.

McNamara P J, Milligan-Monroe K C, Khalili S. Identification, cloning, and initial characterization of rot, a locus encoding a regulator of virulence factor expression in Staphylococcus aureus. *J. Bacteriol.* (2000) 182:3197–3203.

Mishra A K, Choi J, Choi S J. Cyclodipeptides: An Overview of Their Biosynthesis and Biological Activity. *Molecules 22 pii*: (2017) E1796. PMID: 29065531.

Otto M, Echner H, Voelter W. Pheromone cross-inhibition between Staphylococcus aureus and Staphylococcus epidermidis. *Infect. Immun.* (2001) 69:1957–1960.

Pan J. Quorum sensing inhibitors: a patent overview. *Expert Opin. Ther. Pat.* (2009) 19:1581–1601.

Park D K, Lee K E, Baek C H, Kim I H, Kwon J H, Lee W K, Lee K H, Kim B S, Choi SH, Kim K S. Cyclo(Phe-Pro) modulates the expression of ompU in Vibrio spp. *J. Bacteriol.* (2006) 188:2214–2221.

Peterson S, Cline R T, Tettelin H, Sharov V, Morrison DA. Gene expression analysis of the Streptococcus pneumoniae competence regulons by use of DNA micro-arrays. *J. Bacteriol.* (2000) 182:6192–6202.

Pollitt E J, West S A, Crusz S A, Burton-Chellew M N, Diggle S P. Cooperation, quorum sensing, and evolution of virulence in Staphylococcus aureus. *Infect. Immun.* (2014) 82:1045–1051.

Pragman A A, Yarwood J M, Tripp T J. Characterization of virulence factor regulation by SrrAB, a two-component system in Staphylococcus aureus. *J. Bacteriol.* (2004) 186(8):2430–2438.

Prasad C. Bioactive cyclic dipeptides. *Peptides* (1995) 16:151–164.

Rhee K H. Cyclic dipeptides exhibit synergistic, broad spectrum antimicrobial effects and have anti-mutagenic properties. *Int. J. Antimicrob. Agents* (2004) 24:423–427 PMID: 15519471.

Ross P L, Huang Y N, Marchese J N, Williamson B, Parker K, Hattan S, Khainovski N, Pillai S, Dey S, Daniels S, Purkayastha S, Juhasz P, Martin S, Bartlet-Jones M, He F, Jacobson A, Pappin DJ. Multiplexed protein quantitation in Saccharomyces cerevisiae using aminereactive isobaric tagging reagents. *Mol. Cell Proteomics* (2004) 3:1154–1169.

Rutherford S T. Bacterial quorum sensing: its role in virulence and possibilities for its control. *Cold Spring Harb. Perspect. Med.* (2012) 2:a012427.

Saïd-Salim B, Dunman P M, McAleese F M, Macapagal D, Murphy E, McNamara P J, Arvidson S, Foster T J, Projan S J, Kreiswirth B N. Global regulation of Staphylococcus aureus genes by Rot. *J. Bacteriol.* (2003) 185:610–619.

Sundaramoorthy R, Fyfe P K, Hunter. WN Structure of Staphylococcus aureusEsxA suggests a contribution to virulence by action as a transport chaperone and/or adaptor protein. *J. Mol. Biol.* (2008) 383:603–614.

Tal-Gan Y, Ivancic M, Cornilescu G, Blackwell H E. Characterization of structural elements in native autoinducing peptides and non-native analogues that permit the differential modulation of AgrC-type quorum sensing receptors in Staphylococcus aureus. *Org. Biomol. Chem.* (2016) 14:113–121.

Tang K. Quorum Quenching Agents: Resources for antivirulence therapy. *Mar. Drugs* (2014) 12:3245–3282.

Thoendel M, Kavanaugh J S, Flack C E. Peptide signaling in the staphylococci. *Chem. Rev.* (2011) 111:117–151.

Vuong C, Saenz H L, Gotz F. Impact of the agr quorum-sensing system on adherence to polystyrene in Staphylococcus aureus. *J. Infect. Dis.* (2000) 182:1688–1693.

Wang B, Zhao A. Activation and inhibition of the receptor histidine kinase AgrC occurs through opposite helical transduction motions. *Mol. Cell* (2014) 53:929–940.

Wang B. Regulation of Virulence in Staphylococcus aureus: Molecular mechanisms and remaining puzzles. *Cell Chem. Bio.l* (2016) 23(2):214–224.

Wang R, Braughton K R, Kretschmer D, Bach T H, Queck S Y, Li M, Kennedy A D, Dorward D W, Klebanoff S J, Peschel A, DeLeo F R, Otto M. Identification of novel cytolytic peptides as key virulence determinants for community-associated MRSA. *Nat. Med.* (2007) 13:1510–1514.

Wei Y, Perez L J, Ng W. L, Semmelhack M F, Bassler B L. Mechanism of Vibrio cholerae autoinducer-1 biosynthesis. *ACS Chem. Biol.* (2011) 6:356–365.

Wiese S, Reidegeld K A, Meyer H E. Protein labeling by iTRAQ: a new tool for quantitative mass spectrometry in proteome research. *Proteomics* (2007) 7:340–350.

Xiao YY, Kuang JF, Qi XN, Ye YJ, Wu ZX, Chen JY, Lu WJ. A comprehensive investigation of starch degradation process and identification of a transcriptional activator MabHLH6 during banana fruit ripening. *Plant. Biotechnol. J.* (2017) 16:151–164.

Yarwood J M, Bartels D J, Volper E M. Quorum sensing in Staphylococcus aureus biofilms. *J. bacteriol.* (2004) 186:1838–1850.

Yarwood JM, McCormick JK. Identification of a novel two-component regulatory system that acts in global regulation of virulence factors of Staphylococcus aureus. *J. Bacteriol.* (2001) 183(4):1113–1123.

Zhang Q, Huang S, Luo H, Zhao X, Wu G, Wu D. Eight-plexiTRAQ labeling and quantitative proteomic analysis for human bladder cancer. *Am. J. Cancer Res.* (2017) 7:935–945.

Zhao J, Li X, Hou X, Quan C, Chen M. Widespread Existence of Quorum Sensing Inhibitors in Marine Bacteria: Potential Drugs to Combat Pathogens with Novel Strategies. *Mar. Drugs* (2019) 17(5). pii: E275.PMID: 31072008.

Zhu H, Sun SW, Li H, Chang A, Liu YC, Qian J, Shen YL. Significantly improved production of Welan gum by Sphingomonas sp. WG through a novel quorum-sensing-interfering dipeptide cyclo(L-Pro-L-Phe). *Int. J. Biol. Macromol.* (2019) 126:118–122.

Chapter 6

The Importance of Endogenous Plant Elicitor Peptide Signals in Response to Biotic and Abiotic Stresses

Mehdi Safaeizadeh[1,*], Nachelli Malpica-López[2†] and Masoud Tohidfar[1,‡]

[1]Department of Cellular and Molecular Biology,
Faculty of Life Sciences and Biotechnology,
Shahid Beheshti University, Velenjak, Tehran, Iran
[2]Zürich-Basel Plant Science Center, Department of Environmental Sciences,
University of Basel, Basel, Switzerland

Abstract

In the innate immunity in plants, the first layer of defense against any pathogen attack is triggered by the perception of pathogen-associated molecular patterns by specific pattern recognition receptors and subsequently induce pattern triggered-immunity (PTI). The endogenous small peptide signals called damage-associated molecular patterns (DAMPs) are generated in response to biotic and abiotic stresses and activate PTI. Recent findings confirmed the importance of Plant elicitor peptides (Peps) to fine-tune the adequate defense responses to diverse biotic and abiotic stresses. It is confirmed that Peps are not redundant, and each member of this family has a specific function. The Peps are identified in different plant species. Considering their importance in

* Corresponding Author's Email: Mehdi.Safaeizadeh@gmail.com, Ma_Safaei@sbu.ac.ir, Mehdi.Safaeizadeh@unibas.ch.
† Co-author's Email: Nachelli.Malpica@gmail.com.
‡ Co-author's Email: m_tohidfar@sbu.ac.ir, gtohidfar@yahoo.com.

In: Advances in Biology. Volume 5
Editor: Charles D. Grant
ISBN: 979-8-89113-441-6

response to biotic and abiotic stresses and recent discoveries of Peps orthologs in other plant species, Peps-triggered immunity has emerged as a highly complex, very dynamic, and interesting field in plant innate immunity. Furthermore, it is possible to use synthetic Peps to induce the immune system against biotic and abiotic stresses in plants and apply these synthetic peptides to induce resistance against microbial infection. Here, the recent finding and ongoing progress on Pep-triggered immunity are presented. Moreover, the recent finding of Peps orthologs in other plant species is presented in detail.

Keywords: Biotic and abiotic stresses, Endogenous peptide signals, Immune receptor, Plant immunology

Introduction

In plants the innate immunity is initiated by the perception of microbes or the the signatures of the microbes by pattern recognition receptors (PRRs) (Safaeizadeh and Boller, 2019; Safaeizadeh et al., 2021). Plants and other multicellular organisms such as mammals have a highly sophisticated system to monitor cellular integrity and to detect the presence of damaged cells (Couto and Zipfel, 2016; Safaeizadeh and Boller, 2019). In plants as well as in mammals, this is based on the recognition of endogenous host derived elicitors, the so-called "Damage-associated molecular patterns" (DAMPs) (Couto and Zipfel, 2016; Safaeizadeh and Boller, 2019). DAMPs perception leads to the induction of similar defense responses same as the perception of microbe-associated molecular patterns (MAMPs), both in plants and in mammals (Couto and Zipfel, 2016; Safaeizadeh and Boller, 2019).

As the well-defined endogenous peptide signal, *At*Pep1 elicitor is a small peptide was isolated from an extract of wounded *Arabidopsis thaliana* leaves, which could activate defense-related genes and also the production of reactive oxygen species called ROS (Yamaguchi and Huffaker, 2011; Safaeizadeh and Boller, 2019). *At*Pep1 is a 23-aa peptide from *A. thaliana* which is derived from a 92-aa precursor proteins encoded within a small gene called *AtPROPEP1* (Huffaker et al., 2006). Recently it was reported that *AtPROPEP1* is inducible by biotic and abiotic stresses (Safaeizadeh and Boller, 2019). Further analysis reported that *AtPROPEP1* gene has seven paralogues which are named *AtPROPEP1 - AtPROPEP8.* Except *AtPROPEP6* which is located on chromosome 2, all the others are located in chromosome 5. Bioinformatics investigations and genome analysis showed that *AtPROPEPs* have orthologs in other plant species including monocots and dicots) (Ruiz et al., 2018, Foix et al., 2021). The discovery of the endogenous peptide signal *At*Pep1 in *A. thaliana* has opened a new field of plant innate immunity and plant immunology research.

Using the *A. thaliana* as the model, subsequent investigations in other plant species leads to identification of AtPeps orthologos in other plant species. Since endogenous peptide elicitors similar to the *At*Pep family have been identified in different species across the plant kingdom, it seems that they have been maintained over evolution (Ruiz et al., 2021). Anyhow, despite many findings of AtPeps orthologs in several plant species, AtPeps orthologs are not identified in many plant species so far. They have many important roles in regulating and balancing the immune system to attack by both pathogens and also herbivores (Safaeizadeh and Boller, 2019).

In this chapter we will explain the major groups of endogenous peptide signals in plants and explain their differences. So far, several classes of plant-derived molecules, which elicit defense responses, have been identified. Endogenous peptide elicitors in plants are classified into different groups based on the structure of their precursor proteins, which include different processing mechanisms to release the active signal (Dressano et al., 2020; Lori et al., 2015). In the next step of this chapter, as most of our information about endogenous peptide signals are derived from the model plant *A. thaliana*, we explain in more detail the endogenous peptide signals in *A. thaliana*.

It is worth notable that the differences between the amino acid sequences of these endogenous peptide signals in different plant families and species indicate the diversity of receptor partners (Foix et al., 2021; Ruiz et al., 2018; Safaeizadeh and Boller, 2019). These receptors that perceive these peptides also show that there is a diversity in processing and also different export mechanisms for activation of these peptide signals in the cell (Foix et al., 2018; Safaeizadeh and Boller, 2019). Therefore, these endogenous peptide signals can be grouped into three major classes.

Peptides from Precursor Proteins without an N-terminal Secretion Signal

The best example of this group is AtPeps family in *A. thaliana*. They are characterized a well-studied family of endogenous peptide elicitors (Figure 1). They are derived from the family of AtPROPEPs, which do not have an N-terminal secretion signal (Huffaker and Ryan, 2007). Interestingly, it has been proposed that AtPeps signaling in *A. thaliana* intensifies or prolongs the stereotypical defense response triggered by MAMPs (Flury et al., 2013). Thus, it seems that DAMPs are important for the fine-tuning of the defense response (Flury et al., 2013; Yamaguchi et al., 2010). Furthermore, recently it is discovered that AtPep1 is released by activation of Metacaspase MC4 upon wounding of the *A. thaliana*. It seems that for the proper activation of AtPep1, Metacaspases have important role (Hander et al., 2019).

Another well studied member of this group is systemin from tomato, which was the first endogenous peptide signal identified in plants (Wang, et al., 2018). Systemin, a peptide with 18 aa residues, induces various defense responses in tomato leaves and cell cultures (Wang et al., 2018). It is formed from the C-terminal domain of a 200-aa precursor protein. Despite the tremendous effort to identify the receptor for systemin, finding the receptor was not successful; however, it has been speculated that the systemin receptor in tomato most likely belongs to the class of leucine- rich repeat receptor like kinases (LRR-RLKs) (Wang et al., 2018). Recent studies showed that systemin is not the only peptide molecule that elicit defense responses (Wang et al., 2018).

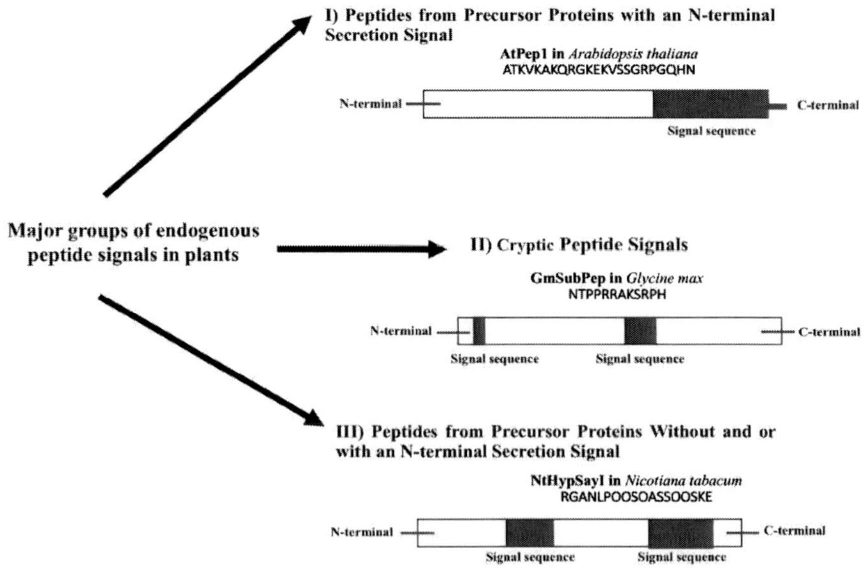

Figure 1. Schematic representation the major groups of the endogenous peptide signals in plants. The best studied example for each group and the amino acid sequence of the active site is illustrated. Precursor protein and peptide sequences for each group are presented. The signal sequence is illustrated in black.

Cryptic Peptide Signals and Their Functions

The terms "cryptic peptides" is used to indicate the pool of peptides formed through the proteolytic action of peptidases on precursor proteins. This group of peptides may have totally different biological activities that can be discriminated from the function of their precursor proteins (Pearce et al., 2010). Formerly, Pearce et al., (2010) identified a 12-aa peptide from soybean which can activate the expression of defense genes upon herbivory attack (Figure 1). Since it is derived from a member of the subtilisin-like protease (subtilase) family, it was named *Glycine max* Subtilase Peptide (GmSubPep). This peptide is regarded as the novel defense-regulating peptide. Perception of GmSubPep leads the alkalinization responses and expression of the defense marker genes. furthermore, perception of GmSubPep by its corresponding receptor leads to the initiation of defense signaling cascades (Pearce et al., 2010). It has been also confirmed that the gene encoding GmSubPep was not induced by defense-related phytohormones or wounding and is constitutively expressed in all actively growing tissues (Bartels and Boller, 2015; Pearce et al., 2010; Safaeizadeh and Boller, 2019; Yamaguchi and Huffker, 2013). It is worth notable that using bioinformatics analysis and in-silico prediction, the GmSubPep-like domain is predicted in *Medicago truncatula.*

Peptides from Precursor Proteins without an N-terminal Secretion Signal

In tobacco two 18-aa glycopeptides induces defense responses (Figure 1). These peptides are named NtHypSysI and NtHypSysII. They are hydroxyproline-rich systemins, and both are derived from the same precursor protein Ntprepro-HypSys (Poretsky et al., 2011), which carries an N-terminal secretion signal. Orthologs of these peptides also have been identified in other solanaceous plants (Poretsky et al., 2011). It is worth notable that in wild tobacco named *Nicotiana attenuata* the endogenous signal peptide is discovered. This peptide is named NaHypSys that regulate the synthesis of anti-herbivory compounds (Heiling et al., 2010). Anyhow, more additional research is needed to identify the novel endogenous peptide signals without N-terminal secretion signals.

Table 1. *AtPROPEP* genes activity in response to Biotic and Abiotic stresses

Biotic stress						Abiotic stress				
Genes name	FLG22	AtPep1	Methyl Jasmonate	Methyl Salicylate	*Pst* DC 3000 infection	Darkness	NaCl	Wounding	Temperature	References
AtPROPEP1	+	+	+	+	+	+	+	+	+	Safaeizadeh and Boller, 2019; Bartels et al., 2013
AtPROPEP2	+	+	+	+	+	+	-	+	+	Safaeizadeh and Boller, 2019; Bartels et al., 2013
AtPROPEP3	+	+	+	+	+	+	+	+	+	Safaeizadeh and Boller, 2019; Bartels et al., 2013
AtPROPEP4	+	+	-	-	+	-	-	-	-	Safaeizadeh and Boller, 2019; Bartels et al., 2013
AtPROPEP5	+	+	+	+	+	-	-	+	+	Safaeizadeh and Boller, 2019; Bartels et al., 2013
AtPROPEP6	+	-	+	-	+	-	-	-	-	Safaeizadeh and Boller, 2019; Bartels et al., 2013
AtPROPEP7	-	-	+	-	+	-	-	-	-	Safaeizadeh and Boller, 2019; Bartels et al., 2013
AtPROPEP8	+	-	+	-	+	-	-	-	+	Safaeizadeh and Boller, 2019; Bartels et al., 2013

Pep Perception Activate the Defense Responses and Re-Program the Cell

As the consequence of Pep perception, several events will occur. Briefly, first of all, ligand binding with AtPEPRs leads to heteromerization with their co-receptor BAK1 (Sun et al., 2013), and subsequently, downstream signaling cascades lead to the release of AtPEPR-bound BIK1 (Choi et al., 2021). Afterward, Ca^{2+}-influx is changed, and as a result, the cytosolic Ca^{2+} levels increase (Qi et al., 2010). This affects the activation of the RbohD protein, which has a crucial role in the oxidative burst, i.e., the formation of ROS (Flury et al., 2013; Krol et al., 2010). In addition, as AtPEPRs contain a cytosolic guanylyl cyclase (GC) domain, the ligand perception may lead to production of cyclic GMP (cGMP) (Ma et al., 2012). In addition, the ROS that are generated may themselves have a role in different defense signaling pathways and also in membrane depolarization (Ruiz et al., 2018). Concomitantly, phosphorylation of a MAPKs mitogen-activated protein kinases (MAPKs), specially MPK3 and MPK6 is occurring (Krol et al., 2010). This phosphorylation may lead to the activation of defense-related transcription factors and ultimately to the induction of many defense-related genes. Also, may lead to an increase in the levels of the defense hormones ET, jasmonic acid and also SA (Yamaguchi, and Huffaker, 2011). After ligand perception and signal transduction, endocytosis and degradation of the receptor may occur, in part via plant U-box type E3 ubiquitin ligases (PUB) mediated processes (Yamaguchi et al., 2010). In addition, in the long term, Pep perception also leads to callose deposition and seedling growth inhibition (Smakowska-Luzan et al., 2018) and also production of secondary metabolites (Tang et al., 2015).

Classification of AtPep Family Based on Their Response to Biotic and Abiotic Stresses

AtPROPEPs classification was a question for researchers for decade. It is needed to classify AtPPROPEPs and their corresponding receptors into several groups. Recent studies showed that *AtPROPEP3* expression is highly induced in response to NaCl treatments (Postel et al., 2010; Safaeizadeh, 2021; Safaeizadeh and Boller, 2019). It seems that they have specific function in response to abiotic stresses. Furthermore, in response to bacterial infection, the expression of *AtPROPEP3* is highly induced (Logeman et al., 2013; Safaeizadeh et al., 2016; Schulze et al., 2010). Recently, Safaeizadeh and Boller (2019) showed that it is possible to subdivide the Pep family into four groups, based on their observation made using promoter-GUS reporter lines in which the promoters of the various *AtPROPEP* genes were fused with the GUS gene. Furthermore, the expression of *AtPROPEP1* gene is highly induced after treatment with *Pseudomonas aeroginosa* (Safaeizadeh and Minae, 2021). Moreover, it is considerable to note that treatment the seedling of the *A. thaliana* with AtPep1 leads the global transcriptional changes compared to the un-treated plants. Interestingly, comparing the level of the global transcriptional changes with the flg22, it was found that AtPep1 treatment increase the rapid transcription of more genes once it was compared with the flg22 treatment (Safaeizadeh et al., 2021).

Recently, Safaeizadeh and Boller (2019) could classify *AtPROPEPs* in different groups based on their expression in response to biotic and abiotic stimuli. They could classify *AtPROPEP1* in one group; *AtPROPEP2* and *AtPROPEP3* in a second group; *AtPROPEP4*, *AtPROPEP7* and *AtPROPEP8* in a third group and *AtPROPEP5* in a fourth group. Their finding, confirm non-redundant roles among the members of the AtPROPEP family and their corresponding receptors. They could also show that among the AtPROPEP family, the AtPROPEP3 is the most active Pep which highly expressed under different stimuli. Furthermore, yellow fluorescent protein (YFP) study which was fused to the AtPROPEP proteins, to determine protein localization showed that AtPROPEP3 was found to be present in the cytosol, while AtPROPEP1 and AtPROPEP6 were observed in tonoplast (Krol et al., 2010). As the AtPROPEPs showed different expression patterns and seemed to be present and active in different regions of the cell, it can be speculated that they do have specific roles and functions.

We have presented the response of *AtPROPEPs* in response to different treatments. As can be seen in the Table 1, *AtPROPEPs* specifically are activated in response to biotic and abiotic stresses. Interesting point is that each member has specific function in response to biotic and abiotic stimuli. As it is illustrated in this Table, *AtPROPEP1* and *AtPROPEP3* are the most active *AtPROPEPs* among other member of this family, while *AtPROPEP4* and *AtPROPEP7* are less active. This finding shows that *AtPROPEPs* are not redundant and each member have specific function in response to biotic and abiotic stresses. Anyhow more additional experiments and more additional treatments are needed to clearly confirm the specific function of *AtPROPEPs*.

Conclusions and Outlook

Peps perception and subsequently signal transduction after Peps perception is a highly conserved mechanisms to rapidly trigger defense responses. The discovery of the AtPeps along with their corresponding receptors, has opened a new field in innate immunity and a novel approach to understand the effect of plant endogenous peptides with regard to their induction upon biotic and abiotic stresses. Safaeizadeh and Boller (2019), clearly showed that the activation of the Pep-family is not redundant and each have a specific function under several forms of biotic and abiotic stresses including MAMP/DAMP treatments, hormone treatments, and salt treatments. In addition, the interplay between MAMP and DAMP signaling has been a question for researchers over several years, and many theories have been proposed so far (as an example the amplifier theory, as proposed by Boller and Felix, (2009).

Although up to now, there are some fragments of information about the activation of the Pep-PEPR system, there is still a lack of information about the regulation of *At*PROPEPs and their corresponding receptors (*At*PEPR1 and *At*PEPR2) in response to biotic and abiotic stresses. Recent studies showed that AtPROPEPs display different expression patterns and also exhibit different localizations, although all of them seem to function in similar ways by inducing defense responses (Bartels and Boller, 2015; Safaeizadeh, and Boller, 2019). Although *AtPROPEPs* was classified into different groups based on promoter reporter lines, more studies are needed to classify them and comprehensively understand the mechanisms underlying DAMP perception. Furthermore, it is recently reported that some of the genes encoding AtPROPEPs are not very active upon MAMP/DAMP elicitors (Bartels and Boller, 2015; Safaeizadeh, and Boller, 2019). Thus, each member of this family has specific functions which can be activated under different circumstances and different condition. Therefore, each AtPROPEPs should be studied in more detail.

Peps together with their corresponding receptors, now emerge as potential targets for plant breeding and novel approach for improving crop immunity. Furthermore, although we have pieces of information in the field of Peps perception in the model plant *A. thaliana* and we have identified their orthologs in several plant species, we still do not know how Pep members in other plant species are perceived. We have a little information about Peps in other plant species. In many plant species, we still could not identify the corresponding receptors of Peps. Therefore, additional research is needed to identify and dissect the role of Peps in other plant species and determine their function in response to biotic and abiotic stresses. Moreover, considering the importance of the endogenous peptide signals in plants, it is suggested to use transgenic plants harboring the genes encoding these endogenous peptides to determine whether transgenic manipulation of endogenous peptide signaling is enhancing resistance to pathogens infection. In addition to that as emerging researches show that the endogenous peptide signals in plants can activate the anti-herbivory compounds in plants, it is suggested to make more additional researches to apply the endogenous peptide signals against herbivory attacks.

Acknowledgments

The author would like to thank Shahid Beheshti university research support. We acknowledge Professor Thomas Boller for his very useful and helpful comments. This research work was supported by Shahid Beheshti University (research number 21.03.99-D-920-1557).

Disclosure of Potential Conflicts of Interest

No potential conflicts of interest were disclosed.

Ethical Approval

This article does not contain any studies with human participants or animals performed by any of the authors.

Funding

This research work was supported by Shahid Beheshti University (research number 21.03.99-D-920-1557).

Consent to Participate

Not applicable.

Consent for Publication

Not applicable.

References

Bartels S., and Boller T. Quo vadis, Pep? Plant elicitor peptides at the crossroads of immunity, stress and development. *J. Exp. Bot* (2015) 66: 5183–5193.

Bartels S., Lori M., Mbengue M., van Verk M., Klauser D., Hander T., Rainer B., Robatzek S., and Boller T. The family of AtPeps and their precursors in Arabidopsis: Differential expression and localization but similar induction of pattern-triggered immune responses. *J. Exp. Bot,* (2013) 64: 5309–5321.

Choi J. H., Oh, E. S., and Oh M. H. Phosphorylation of BIK1 is critical for interaction with downstream signaling components. *Genes Genomics* (2021) 43(11): 1269-1276.

Couto D., and Zipfel C. Regulation of pattern recognition receptor signaling in plants. *Nat. Rev. Immunol* (2016) 16: 537–552.

Dressano K., Weckwerth P. R., Poretsky E., Takahashi Y., Villarreal C., Shen Z., Schroeder J. I., Briggs S. P., and Huffaker A. Dynamic regulation of Pep-induced immunity through post-translational control of defence transcript splicing. *Nat Plants* (2020) 6(8): 1008-1019.

Flury P., Klauser D., Schulze B., Boller T., and Bartels S. 2013. The anticipation of danger: microbe-associated molecular pattern perception enhances AtPep-triggered oxidative burst. *Plant Physiol* (2013) 161: 2023–35.

Foix L., Nadal A., Zagorščak M., Ramšak Ž., Esteve-Codina A., Gruden K., and Pla M. Prunus persica plant endogenous peptides PpPep1 and PpPep2 cause PTI-like transcriptome reprogramming in peach and enhance resistance to Xanthomonas arboricola pv. pruni. *BMC Genomics* (2021) 18;22(1):360.

Hander T., Fernández-Fernández Á. D., Kumpf R. P., Willems P., Schatowitz H., Rombaut D., Staes A., Nolf J., Pottie R., Yao P., Gonçalves A., Pavie B., Boller T., Gevaert K., Van Breusegem and F., Bartels S., and Stael S. Damage on plants activates Ca^{2+}-dependent metacaspases for release of immunomodulatory peptides. *Science* (2019) 22;363(6433).

Heiling S., Schuman M. C., Schoettner M., Mukerjee P., Berger B., Schneider B., Jassbi A. R., Baldwin I. T. Jasmonate and ppHsystemin regulate key malonylation steps in the biosynthesis of 17-hydroxygeranyllinalool diterpene glycosides, an abundant and effective direct defense against herbivores in *Nicotiana attenuata*. *Plant Cell* (2010), 22:273-292.

Huffaker A., and Ryan C. A. Endogenous peptide defense signals in Arabidopsis differentially amplify signaling for the innate immune response. *Proc. Natl Acad. Sci. USA* (2007) 104:10732–10736.

Huffaker A., Pearce G., and Ryan C. A. An endogenous peptide signal in Arabidopsis activates components of the innate immune response. *Proc. Natl Acad. Sci. USA* (2006) 103: 10098–10103.

Huffaker A., Pearce G., Veyrat N., Erb M., Turlings T. C., Sartor R., Shen Z., Briggs S. P., Vaughan M. M., Alborn H. T., Teal P. E., and Schmelz E. A. Plant elicitor peptides are conserved signals regulating direct and indirect antiherbivore defense. *Proc. Natl Acad. Sci. USA* (2013) 110: 5707–5712.

Krol E., Mentzel T., Chinchilla D., Boller T., Felix G., Kemmerling B., Postel S., Arents M., Jeworutzki, E., Al-Rasheid K. A., Becker D., and Hedrich R. Perception of the *Arabidopsis* danger signal peptide 1 involves the pattern recognition receptor *At*PEPR1 and its close homologue *At*PEPR2. *J. Biol. Chem* (2010) 285: 13471–13479.

Logemann E., Birkenbihl R. P., Rawat V., Schneeberger K., Schmelzer E., and Somssich I. E. Functional dissection of the PROPEP2 and PROPEP3 promoters reveals the importance of WRKY factors in mediating microbe-associated molecular pattern-induced expression. *New Phytol* (2013) 198: 1165–1177.

Lori M., van Verk M. C., Hander T., Schatowitz H., Klauser D., Flury P., Gehring C. A., Boller T., and Bartels S. 2015. Evolutionary divergence of the plant elicitor peptides (Peps) and their receptors: interfamily incompatibility of perception but compatibility of downstream signalling. *J Exp Bot* (2015) 66(17): 5315-25.

Ma Y., Walker R. K., Zhao Y., and Berkowitz G. A. Linking ligand perception by PEPR pattern recognition receptors to cytosolic Ca2+ elevation and downstream immune signaling in plants. *Proc. Natl Acad. Sci. USA* (2012) 109: 19852–19857.

Pearce G., Yamaguchi Y., Barona G., and Ryan C. A. A subtilisin-like protein from soybean contains an embedded, cryptic signal that activates defense-related genes. *Proc Natl Acad Sci U S A* (2010) 107(33):14921-5.

Poretsky E., Dressano K., Weckwerth P., Ruiz M., Char S. N., Shi D., Abagyan R., Yang B., and Huffaker A. Differential activities of maize plant elicitor peptides as mediators of immune signaling and herbivore resistance. *Plant J* (2020) 104(6): 1582-1602.

Postel S., Küfner I., Beuter C., Mazzotta S., Schwedt A., Borlotti A., Halter T., Kemmerling, B., and Nurnberger T. The multifunctional leucine-rich repeat receptor kinase BAK1 is implicated in Arabidopsis development and immunity. *Eur. J. Cell Biol,* (2010) 89: 169–174.

Ruiz C., Nadal A., Foix L., Montesinos L., Montesinos E., and Pla M. Diversity of plant defense elicitor peptides within the Rosaceae. *BMC Genet* (2018) 23;19(1):11.

Ruiz C., Nadal A., Montesinos E., and Pla M. Novel Rosaceae plant elicitor peptides as sustainable tools to control Xanthomonas arboricola pv. pruni in Prunus spp. *Mol Plant Pathol* (2018) 19(2): 418-431.

Safaeizadeh M., and Boller T. Differential and tissue-specific activation pattern of the AtPROPEP and AtPEPR genes in response to biotic and abiotic stress in Arabidopsis thaliana. *Plant Signal Behav* (2019) 14, e1590094.

Safaeizadeh M., and Minae D. Study the effect of oil extracted *Pseudomonas aeruginosa* to stimulate immune system in *Arabidopsis thaliana. Ninth International Congress on Biology, Semnan*, Iran (2021) 21: 0796.

Safaeizadeh M., Bartels S., and Boller T. Regulation of the *AtPROPEP* and *AtPEPR* Genes under Biotic and Abiotic Stress, Studied with Promoter-GUS-reporter Constructs in *Arabidopsis thaliana. Epigenetic and Chromatin Regulation Congress Strasburgh*, France (2016) P.51.

Safaeizadeh M., Boller T., and Becker C. Transcriptomic profiling uncovers novel players in innate immunity in *Arabidopsis thaliana* Cold Spring Harbor. Biorxiv. (2021) 01.02.425067.

Safaeizadeh M. Study the effect of salt treatment on the expression of the AtPropep family in Arabidopsis thaliana. *9th International Congress on Biology*, Semnan, Iran. (2021) 21: 0797.

Schulze B., Mentzel T., Jehle A. K., Mueller K., Beeler S., Boller T., Felix G., and Chinchilla D. Rapid heteromerization and phosphorylation of ligand-activated plant transmembrane receptors and their associated kinase BAK1. *J. Biol. Chem* (2010) 285: 9444–9451.

Sun Y., Li L., Macho A. P., Han Z., Hu Z., Zipfel C., Zhou J., and Chai J. Structural Basis for flg22-Induced Activation of the *Arabidopsis* FLS2-BAK1 Immune Complex. *Science* (2013) 6158: 624–628.

Tang J., Han Z., Sun Y., Zhang H., Gong X., and Chai J. Structural basis for recognition of an endogenous peptide by the plant receptor kinase PEPR1. *Cell Research* (2015) 25: 110–120.

Wang L., Einig E., Almeida-Trapp M., Albert M., Fliegmann J., Mithöfer A., Kalbacher H., and Felix G. The systemin receptor SYR1 enhances resistance of tomato against herbivorous insects. *Nat Plants* (2018) 4(3): 152-156.

Yamaguchi Y., and Huffaker A. Endogenous peptide elicitors in higher plants. *Curr. Opin. Plant Biol* (2011) 14: 351–357.

Yamaguchi Y., Huffaker A., Bryan A. C., Tax F. E., and Ryan C. A. PEPR2 is a second receptor for the Pep1 and Pep2 peptides and contributes to defense responses in Arabidopsis. *Plant Cell* (2010) 22: 508–522.

Biographical Sketch

Mehdi Safaeizadeh

Affiliation: Assistant Professor.

Education:

PhD from the University of Basel under the direct supervision of Professor Thomas Boller.

Business Address:

Department of Cellular and Molecular Biology, Faculty of Life Sciences and Biotechnology, Phone number: +98(0)91/2037-3610, Shahid Beheshti University, Postal Code: 1983969411, Velenjak, Tehran, Iran.

Research and Professional Experience:

Plant molecular biology, plant microbe interaction, innate immunity, signaling.

Professional Appointments:

Honors:

Ranked one among all graduated students during academic studies and top researcher as an assistant professor.

Publications from the Last 3 Years:

[1] Safaeizadeh, M. Saeidi, A. Fallah M. (2023). Bioinformatic investigation of proteinstability CMV Cucumber mosaic virus. *10th International Congress on Medicinal Plants. p.* 279. Urmia, Iran.

[2] Safaeizadeh, M. Ghotbi-Ravandi, A. A., and Fallah, M. (2023). Investigation of hormonal changes in Arabidopsis thaliana plant under the effect of biological stress by Pseudomonas aeruginosa. *10th International Congress on Medicinal Plants*. p. 280. Urmia, Iran.

[3] Safaeizadeh, M. Amini Neisiani, A. and Saedi, A. (2023). Investigating the morphological effect of salicylic acid treatment on resistance to pathogenic bacteria in *Nicotiana Rustica. 10th International Congress on Medicinal Plants.* p. 281. Urmia, Iran.

[4] Safaeizadeh, M. Fallah Ziarani, M., Altafian, Y. and Ghotbi-Ravandi, A. A. (2023). Bioinformatic investigation of protein stability Calmodulin-binding Protein 60 g (CBP60g) Cucumber mosaic virus. *10th International Congress on Medicinal Plants. p.* 283. Urmia, Iran.

[5] Safaeizadeh, M. Bamdad, F. Fallah, M. and Darvishi, F. (2023). Investigation the effect of Saccharomyces cerevisiae inoculation on Nicotiana tabacum plant hormones. *5th international and 17th Iranian Genetic Congress*. 71444.

[6] Safaeizadeh, M. Bamdad, F. Fallah, M. and Jahanfar, F. (2023). Investigation the effect of treatment of Pseudomonas syringae, Pseudomonas aeroginosa and Saccharomyces cerevisiae on Arabidopsis thaliana using electron microscopy. *5th international and 17th Iranian Genetic Congress*. 71443.

[7] Safaeizadeh, M. Bamdad, F. Fallah, M. and Darvishi, F. (2023). Investigating the effect of Pseudomonas arginosa inoculation on Nicotina tabacum plant hormones. *5th international and 17th Iranian Genetic Congress.* 71437.

[8] Safaeizadeh, M. Bamdad, F. Fallah, M. and Darvishi, F. (2023). Investigating the effect of Pseudomonas syringae on Niconiana tabacum plant hormones. *5th international and 17th Iranian Genetic Congress.* 71441.

[9] Safaeizadeh, M. Boller, T. and Becker, C. (2023). *Comparative RNA-seq analysis of Arabidopsis thaliana response to AtPep1 and flg22, reveals new players in patterntriggered immunity*. Revised.

[10] Safaeizadeh, M. and Fallah, M. (2023). In Silico analysis of 6-fructose transferase protein in Triticum aestivum. *Scientific Quarterly Journal of Crop Plant Biotechnology*. Accepted.

[11] Safaeizadeh, M. (2022). Endogenous Peptide Signals in Arabidopsis thaliana, their receptors and their role in innate immunity. *Review. J Plant Mol Breed*. 9 (1): 1-11.

[12] Safaeizadeh, M. and Fallah, M. Bamdad, F. Darvishi, F. (2022). Evaluation of the stability of the immune Receptor Flagellin Sensitive-2. *6th International Congress of Biomedical research* (ICB2022).

[13] Safaeizadeh, M. and Fallah, M. Bamdad, F. Darvishi, F. (2022). In silico analysis of the second and third structure of the immune receptor Flagellin Sensitive-2. *6th International Congress of Biomedical research* (ICB2022).

[14] Fallah, M. Safaeizadeh, M. and M. Bamdad, F. (2022). Bioinformatic analysis of the secondary and tertiary structure of protein 6-fructose transferase in wheat. *22th National and 10th International Congress on Biology*. (06)1143.

[15] Bamdad, F. Safaeizadeh, M. and Fallah, M. (2022). Investigating the effect of oil-eating bacteria Pseudomonas androgenosa on the change of hormones in the model plant *Nicotiana benthamiana*. *22th National and 10th International Congress on Biology*. 06(1082).

[16] Safaeizadeh, M. (2021). Study the effect of NaCl stress on the expression of the AtPropep family in *Arabidopsis thaliana*. *21th National and 9th International Congress on Biology*. 283 (99)1462.

[17] Safaeizadeh, M. and Minae, D. (2021). Study the effect of oil extraced *Pseodomonas aeruginosa* to stimulate immune system in *Arabidopsis thaliana*. *21th National and 9th International Congress on Biology*. 283 (99)1408.

[18] Safaeizadeh, M. (2022). First report of cucumber mosaic virus infecting Dianthus hybridus in Iran. *Journal of Plant Pathology*. p.397. https://doi.org/10.1007/s42161-021-00958-3.

[19] Safaeizadeh, M. and Malpica, N. (2021). In silico analysis and molecular investigation of the nuclear shuttle protein-interacting kinase 1 (NIK1) in six Solanaceous species. *Cold Spring Harbor. Biorxiv*. https://doi.org/10.1101/2021.07.06.451341.

[20] Safaeizadeh, M. Boller, T. and Becker, C. (2021). Transcriptomic profiling uncovers new players in innate immunity in Arabidopsis thaliana. *Cold Spring Harbor molecular. Biorxiv*. https://doi.org/10.1101/2021.01.02.425067.

[21] Safaeizadeh, M. and Boller T. Differential and tissue-specific activation pattern of the AtPROPEP and AtPEPR genes in response to biotic and abiotic stress in Arabidopsis thaliana. *Plant Signal Behav*. 2019;14(5):e1590094. doi: https://doi.org/10.1080/15592324.2019.1590094. Epub 2019 Mar 24. PubMed PMID: 30907222; PMCID: PMC6512929.

Chapter 7

The Mechanism of Influence of Celestial Bodies on Activity and Interaction of Domains (A- and B-Tubulins) with Their Surroundings

Anatoly Zubow[2]
Kristina Zubow[1]
and Viktor Zubow[1]
[1]Zubow Consulting, NPO, Germany
[2]Group of the Telecommunication Networks Division, TU, Berlin, Germany

Abstract

The effect of gravitational fields of gas giants in the solar system on the oscillations of α- and β-tubulins has been studied. The influence of planets on hydrogen bridges (HBs) between tubulins and their surroundings has been established. The mechanism of this influence is proposed, including the influence of planets on the processes of equilibrium of baryon protons of the HB with their forms dissolved in the physical vacuum (PV). A correlation was found between the masses of tubulins in their native state in human skin cells and the equilibria of protons with PV, the influence of planets. The influence of the proton-PV balance shift on tubulin conformations and cell divided processes (grain yield) is discussed.

Keywords: tubulin, mass, hydrogen bridges, proton hopping, grain yields, lean, celestial bodies

In: Advances in Biology. Volume 5
Editor: Charles D. Grant
ISBN: 979-8-89113-441-6

1. Introduction

Domains, in the structural organization of non-crystalline systems, play an important role in the construction and properties, for example, the biological structures of the cell (cytoskeleton) and their division, the transfer of information flows (nanotubes) and their control.

Consider the domains of α- and β-tubulin (Figure 1), it is believed that they exist in the form of dimers with a mass of 55 kDa each, although there are reports of tubulins with other masses, for example 50 kDa (https://ru.wikipedia.org/ wiki/TUBB8). Modern methods for determining the molecular masses of tubulins are not perfect and do not give an understanding of their native, not distorted by rigid methodology, structure. Such domains interact with their surroundings through hydrogen bonds, which are conventionally shown in Figure 1 by dotted lines. The strength of hydrogen bonds varies from 6 to 160 kJ/mol and depends on the polarity of the complex. However, there is no convincing explanation for this difference.

It is reasonable to believe that the tubulin constructions and their quality will depend to a large extent on the quality of the hydrogen bonds of the monomers with their surroundings. On the other hand, there are a number of hydrogen bridges that differ both in number and in bond energies [1, 2]. The hydrogen bridges themselves in physical chemistry, and in solid state physics as well, represent, for example, for two water molecules, as a certain complex $(H_2O{\bullet}H{\bullet}OH_2)^+$, "turning a blind eye" to the absurdity of the scheme involving the proton (H), the role and the nature of which is not clear here.

Another problem arose, namely, tubulins, during the observation period in 2010 ± 3, were registered as physical oscillators with masses of 55 kDa, and in 2020 ± 3 already as with masses of 46 kDa. The reason for this phenomenon remained unclear. Understanding the causes of changes in tubulin activity during cell division is very important [3]. Therefore, we turned our attention to the study of this phenomenon.

The aim of this work was to understand the phenomenon of different masses of tubulin, as an oscillator in the native state, directly in the cell of the body.

2. Material and Method

The object of the study was the dermis of the skin flap of the heel of probant-57 (57 years old) and his own at 70 years old, Figure 4. Gravitational mass spectroscopy (GMS) was used as a non-invasive diagnostics of the cytoskeleton and long-range molecular order in the skin. The GMS sensor touched the surface of the skin flap or was inserted into it to a depth of 0.1 mm, and gravitational noise (GN) from oscillations of atomic nuclei clusters (ANC) in tubulins at 1471 Hz was recorded. GMS spectra were obtained using the first Zubow equation [1]. Positive values of Δf are typical for loose ANC conformations, they reflect the energy fraction of the molecular bunch in the entire ensemble of studied masses from 200 Da to 4 billion Da, negative $-\Delta f$ values reflect the same, but for dense ANC conformations. Δf is the result of subtracting from the gravitational noise (GN) spectrum in the natural background GN sample. Samples (average of 10 measurements) were maximally protected from external energy physical influences (electromagnetic radiation, mechanical fields and strong sudden pressure beads). The method is based on the analysis of GN from the proton jumping in the ANC from the baryonic state to the state of protons in the physical vacuum, Figure 5. Super positions of oscillations of proton ensembles in the ANC (https://translated.turbopages.org/proxy_u/en-ru.ru.e9dfd034-64abf f47-bc78e2b9-74722d776562/https/en.wikipedia.org/wiki/Chladni%27s_ law) allow you to get an idea of their masses and interactions of the ANC with the environment. Scan time 10 sec. The spectra were calibrated by the masses of water clusters [2], nucleosomes with one DNA loop (https://ru.wikipedia. org/wiki/ Nucleosome, 233,599 Da), human mitDNA rings (10,653,867 Da, 16,569 bp, https: //ru.wikipedia.org/wiki/ Mitochondrial_DNA) and mitDNA rings in sugar beet [4] (237,137,757 Da). The accuracy of recording the oscillator masses is no worse than 0.01%.

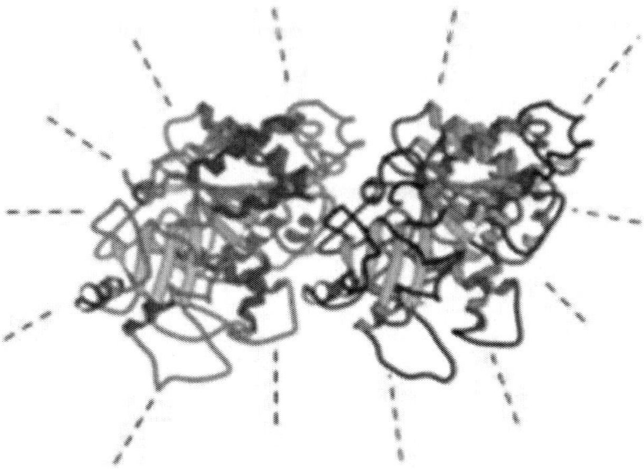

Figure 1. α- and β-tubulin dimer (https://ru.wikipedia.org/wiki/Tubulin), molecular weight 55 kDa and a model of its interaction with its surroundings via hydrogen bonds (https://ru.wikipedia. org/wiki/Hydrogen_bond).

3. Analysis and Discussion

The position of the planets on 26.7.2010 and 26.7.2023 can be found on the website: https://liveplanets.ru/tools/SolarSystem/. Models of their gravitational fields are given in Figure 2.

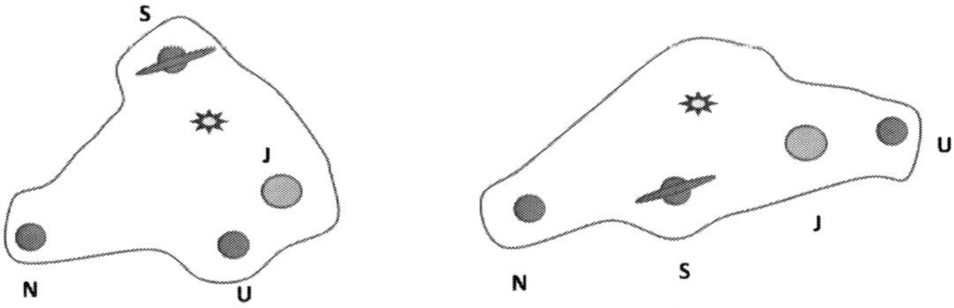

Figure 2. Models of the gravitational fields of gas giants in 2010 and 2023. **N** is Neptune, **S** is Saturn, **J** is Jupiter and **U** is Uranus. The lines outline the super positions of the hypothetical gravitational fields of the gas giants.

Figure 3 shows models of the interaction of α-tubulin with its surroundings at different times of observation. In 2010, the oscillator mass coincided with the literature data with very high accuracy. At this time, the interaction of the domain with its surroundings was weakened, and in 2023 it significantly increased and the mass of the oscillator fell by 16%. Loose forms began to dominate in tubulin conformations. This means that the activity of tubulins, their conformations depend not only on changes in their surroundings, but also on the quality of hydrogen bridges. Let us consider the state of hydrogen bridges in more detail.

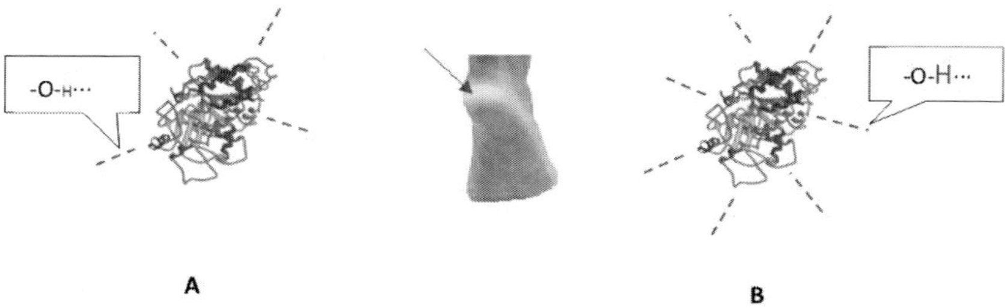

Figure 3. Models of tubulins (for example, α-) as oscillators in the network of interaction with their surroundings for: **A** - 2010 (55 kDa) and **B** - for 2023 (46 kDa). Hydrogen bonds of the domain with its surroundings are indicated by dotted lines. See Figure 6 for explanation.

In 2010, we discovered the phenomenon of proton dissolution in the hydrogen bridges (HB in water) in a physical vacuum (PV) when a hydrogel sample hits the opposition of two gas giants, Saturn and Jupiter [2, 5]. In this regard, it was reasonable to believe that a change in the gravitational fields formed by gas giants around the star and minor planets affects the equilibrium: the baryon proton of the hydrogen bridge and the state of the dissolved form of the proton in the PV (Figure 3), namely, the weakening or strengthening interactions of the domain with its surroundings. According to Figure 3, strong HBs arose in 2023, when the conditions for the dissolution of protons in PV were weakened.

The dissolution of protons in PV [6] has a multistage character. First, protons dissolve to neutrinos and then they dissolve to preons, and the lower the gravitational field strength, the greater the dominance of the baryonic proton over its form dissolved in the PV. It follows from this that when opposed to gas giants in hydrogen bridges, the proton dissolves completely to preons, and under other conditions in the HB this process is less pronounced and the proton is predominantly in the baryonic form, and in solution in the PV only up to its constituent neutrinos, Figure 4. From this it follows that the frequency with which the proton oscillates in the HB between its two forms is not sinusoidal, but pulsed, not uniform, Figure 5.

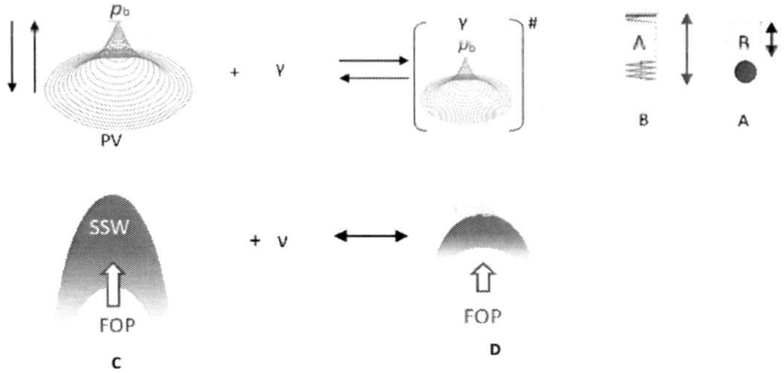

Figure 4. Model of proton oscillations between the physical vacuum and its baryon state (p_b), and the absorption-desorption equilibrium of the interaction of a proton with a neutrino (ν). For understanding, the oscillation loading models are given on the right and below. **A** – unloaded and **B** – oscillations loaded with adsorbed neutrinos. The so-called "pancakes" (**C** and **D**, https://www.youtube.com/watch?v =l yIEBm5u7-Q) have a vector (broad arrow - gravitational compass) in the direction of a larger bunch of baryonic masses or regions of the surrounding space with a higher FOP content. SSW is a standing shock wave from a neutrino (halo) around a proton. PV - designation of physical vacuum.

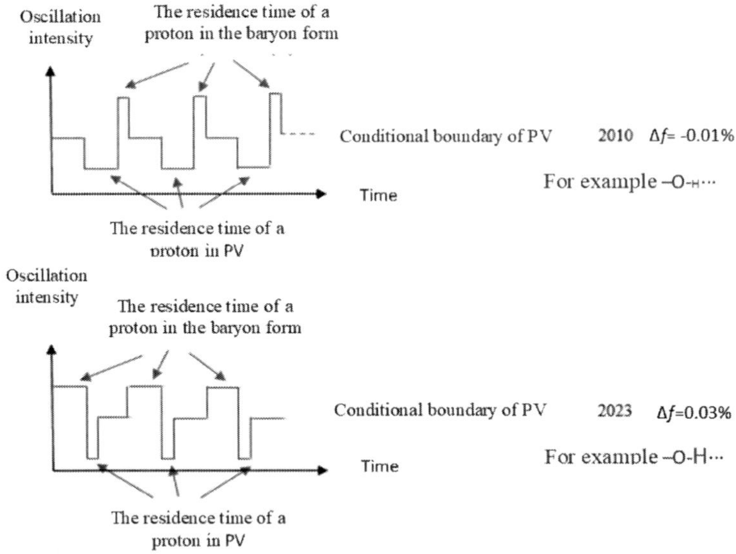

Figure 5. Models of oscillations (hypothetical) of a proton in HB between its baryonic form and the form dissolved in PV in 2010 and 2023 (in tubulin).

Here it is reasonable to believe that in weakened gravitational fields, tubulin conformations will be dominated by active, loose forms with a strong interaction with their surroundings. In this case, the processes of cell division should be accelerated, but each acceleration reduces the assembly quality, so one would expect a negative effect of celestial bodies on the processes of mitosis, including all its stages, and hence an increase in the intensity of mutagenesis.

Data on grain yields in the Russian Federation in 2010 and 2023 (150 million tons) can confirm the above conclusion [7].

Note that in the lean years of 1932/33 (https://pontokot.livejournal.com/719213.html), the constellation of gas giants was close to that in 2010 (Figure 2), see Figure 6. It can be assumed that here, too, the gravitational fields of gas giants played an important role in influencing the active conformations of tubulins.

Conclusion

The mass of the native, active form of tubulins, as oscillators in cells, depends on the state of gravitational fields in the solar system.

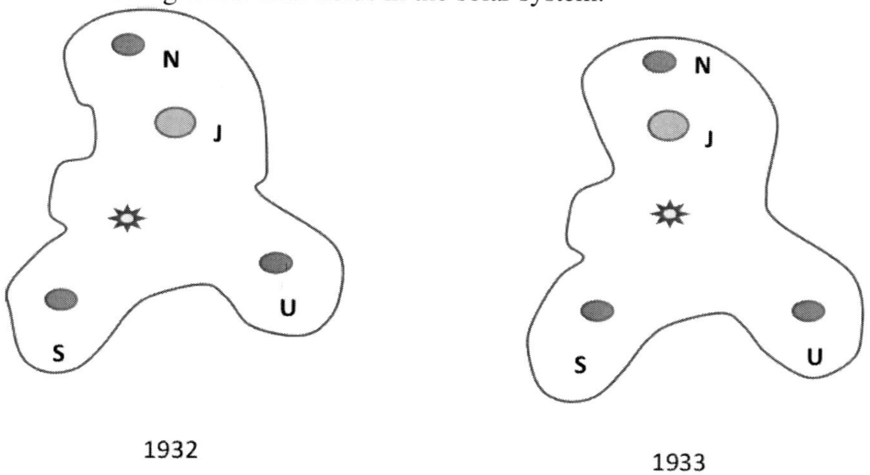

Figure 6. Models of the gravitational fields of gas giants in the lean years of 1932 and 1933. **N** is Neptune, **S** is Saturn, **J** is Jupiter and **U** is Uranus. The lines outline the superpositions of the hypothetical gravitational fields of the gas giants.

The mass of the native, active conformation of tubulins, as oscillators, can become a sensor and indicator of the state of gravitational fields and their dynamics in the Universe.

The quality and energy characteristics of protons in hydrogen bridges between tubulins and their surroundings can be used to monitor gravitational fields and the state of the physical vacuum in the Universe.

The high-frequency (~5E-22 s) balance of baryon protons in hydrogen bridges with their forms in the physical vacuum is not sinusoidal, but is expressed in complex sequences of dissolution and condensation of protons, passing through the stages of decay into smaller elementary particles and *vice versa* during the condensation of a proton from physical vacuum.

There is a fundamental possibility of external influence on the equilibrium of protons with the physical vacuum, and hence the influence on the physicochemical properties of domains and their surroundings.

Atomic nuclei clusters in tubulins are formed by the gravitational noise of the Universe.

Sponsoring

To continue work in this direction, we are looking for sponsors. Write to us.

References

[1] Lee, Han-Myoung. Theoretical Study of the Lowest Energy Structure of the Water Undecamer. *Bulletin of the Korean Chemical Society*. 2003, v. 24, no. 6, pp.777-779. https://doi.org/10.5012/bkcs.2003.24.6.777 .

[2] Zubow K., Zubow A. V., Zubow V. A. The Way to the ETIs. *Applied gravitational mass spectroscopy*. Nova Sci. Publ. NY, 2014. https://www.novapublishers.com/catalog/product_info.php?products_id=42668&osCsid=5bd85d42dc273360fd4812 6de7be9daf .

[3] Zubow K., Zubow A.V., Zubow V.A. Ensemble of tubulins in onion root and their properties. Russian agricultural science. 2018, v.6. pp. 21-25. RUSSIAN AGRICULTURAL SCIENCES. Allerton Press, Inc. 2019, v. 45, pp. 30-34. ISSN 1068-3674, Russian Agricultural Sciences, 2019, Vol. 45, No. 1, pp. 30–34. © Allerton Press, Inc., 2019. Published in Russian in Rossiiskaya *Sel'skokhozyaistvennaya Nauka*, 2018, No. 6, pp. 21–25, doi: 10.3103/S10683674 19010221.

[4] Zubow K., Zubow A., Zubow V. A. *How Nuclear DNA Affects Mitochondrial DNA Horizons in World Physics*. Ed. Albert Reimer. Nova Publisher Inc. NY. 2023,

volume 308, pp. 185-196 (in Book, 140-156). Horizons in World Physics. Volume 308 – Nova Science Publishers (novapublishers.com).

[5] [5]. Zubow K., Zubow A.V., Zubow V.A. The Phenomenon of Proton Dissolving in Vacuum and of Proton Condensation from Vacuum. Two Forms of Protons, Structure of Nuclei, Electrons and Atoms. *J. of modern physics*, vol.1, no.1, 2010 pp.175-184.
https://www.researchgate.net/publication/228928173_The_Phenomenon_of_Proton_Dissolving_in_Vacuum_and_of_Proton_Condensation_from_Vacuum_Two_Forms_of_Protons_Structure_of_Nuclei_Electrons_and_Atoms.

[6] Zubow K., Zubow A., Zubow V. New constant Q=8.97082E-10, $m^2/s^2 \cdot kg^3$, In $E=Q \cdot m^4$,Vector nature of masses. *Horizons in World Physics*. Ed. Albert Reimer. NY, 2023, vol. …, pp.. in print.

[7] (https://yandex.com/images/search?img_url=https%3A%2F%2Fxn--80apggvco.xn--p1ai%2Fimg%2Ftablica-urozhajnost-zernovyh-i-zernobobovyh-v-rossii.jpg&lr=10000&pos=5&rpt=simage&source=serp&text=урожайность%20зерновых%20в%20мире%20по%20годам%20таблица).

Chapter 8

Naturally Occurring Thiolane-Type Cyclic Sulfides: A Review on Mode of Action

Idris Arslan*
Zonguldak Bülent Ecevit University, Biomedical Engineering, Incivez, Zonguldak, Turkey

Abstract

One of the main efforts of researchers to study drug development and design is the discovery of novel agents which could be beneficial in the treatment of diseases in humans. Natural products are complex secondary metabolites which are synthesized by different natural sources such as plants, animals, bacteria and fungi to optimize nature.

This book chapter highlights the naturally occurring thiolane type cyclic sulfides and their therapeutical potency particularly anticancer activity. Thiolane-type cyclic sulfides are especially synthesized by *Allium* species.

Keywords: allium, thiolane, cyclic sulfides, anticancer

Introduction

In spite of the significant improvements in drug design over the past decades, certain disorders including cancer, inflammatory, and infectious diseases continue to pose a great threat to human health. Natural products are complex

* Corresponding Author's Address: Bülent Ecevit University, Biomedical Engineering, 67100 Incivez, Zonguldak, Turkey, Email: idris.arslan@beun.edu.tr or idris.arslan@yahoo.com, https://orcid.org/0000-0001-7634-7148.

In: Advances in Biology. Volume 5
Editor: Charles D. Grant
ISBN: 979-8-89113-441-6

metabolites synthesized by different sources to optimize nature. They have recently gained significant attention in this regard due to their pharmacological potency.

Garlic is an underground bulb belonging to the *Allium sativum* (Liliaceae family). It has been long favored for as a seasonal vegetable including the whole plant, garlic leaf, flower bolt and folk medicine throughout the world. Recently, there has been a considerably increase in the cultivation and quality of garlic, with the garlic production area surpassing 5.2 million hectares all over the world (FAO 2021).

Figure 1. *Allium sativum* (garlic) includes cyclic sulfide compounds.

So far, garlic has been reported to contain many complex phytochemicals such as sulfur-containing compounds, flavonoids, vitamins, saponins, and polysaccharides. Sulfides from *Allium* species began with the purification of allicin and alliin compounds in 1944 and 1951, respectively (Cavallito and Bailey 1944; Stoll and Seebeck 1951).

A cyclic sulfide could be defined as the compounds in which sulfur atom is simultaneously connected to different positions of the same carbon chain. Sulfur belongs to the same group in the periodic table as oxygen but is much less electronegative. This distinctive difference clarifies several distinguishing characteristics of the cyclic sulfide compounds.

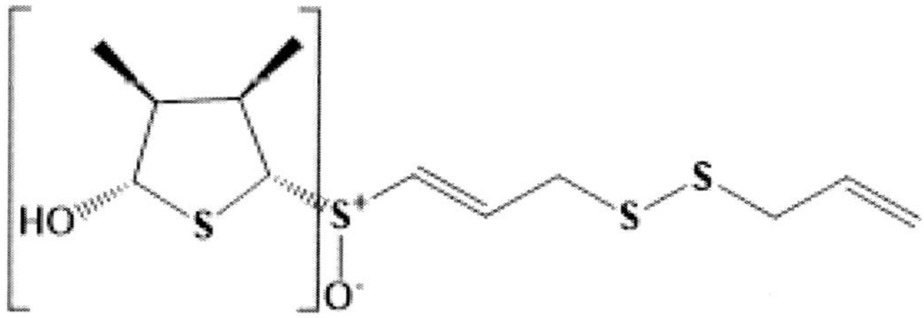

Garlicnins A

Occurrence, Identification, and Biological Activity

Humanity has used, for thousands of years, many cyclic sulfide-containing plants such as garlic, onion and welsh onion (Nohara et al., 2021). In a recent study, Nohara's group from Japan has revealed the isolation, structural elucidation and pharmacological actions of both cyclic and non-cyclic sulfides from several *Allium* species (Gao et al., 2023). Results displayed that acetone was a good solvent for extraction of cyclic sulfides by creating a stable medium for allyl sulfenic acid and allyl thio-sulfenic acid, which were derived from the seperation of allicin (Gao et al., 2023; Gao et al., 2021). Lately, nuclear magnetic resonance (NMR) spectroscopy are combined with Mosher method for structural elucidation (Mosher et al., 1973a/b) due to difficulty in crystalization of some cyclic sulfides.

Cepathiolane is the first plant-derived thiolane compound with 5-membered heterocycle carrying one sulfur atom and isolated from Welsh onion by Yoshida and colleagues. Due to its flavor and Welsh onion scent, cepathiolane compound was patented (Yoshida et al., 2010).

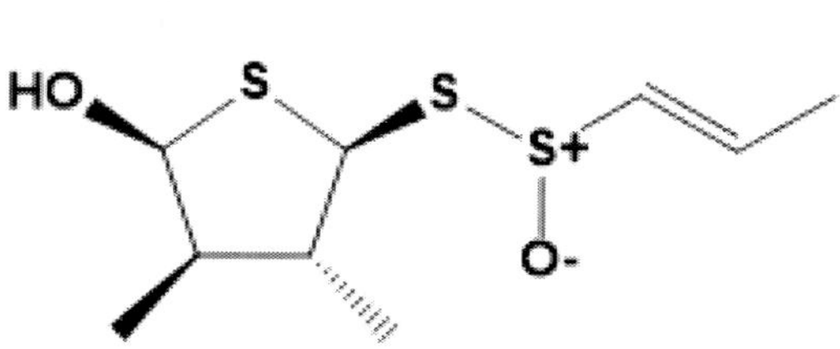

Cepathiolane

Onionin A_1 was first isolated from *Allium cepa* by El-Aasr et al., 2010 as 3,4-dimethyl-5-(1*E*-propenyl)-tetrahydrothiophen-2-sulfoxide-*S*-oxide but Stefano et al., 2019 corrected its structure as (*E*)-3,4-dimethyl-5-(1-propenylsulfinyl) thiolane-2-ol. Furthermore, the stereoisomers onionin A_2 and A_3 were simultanously identified as 3,4-dimethyl-5-(1*E*-propenyl)-tetrahydrothiophene 2-sulfenic acid-*S*-oxides from *Allium cepa* and *A. fistulosum* (Welsh onion) (Nohara et al., 2014). Regarding the biological activities of cyclic sulfides, onionin A_1 has been extensively examined in both *in vitro* and/or *in vivo* cytotoxicity assays. Tsukobi et al., 2016 revealed onionin A_1 reduced the extent of ovarian cell proliferation induced by co-culture with human macrophages. Additionally, onionin A_1 directly blocked cancer cell proliferation. It was also reported that a combinatorial effect with onionin A_1 and anti-cancer drugs. The activation of signal transducer and activator of transcription 3 (STAT3) involved in cell proliferation and chemo-resistance was remarkable inhibited by onionin A_1 in ovarian cancer cells. Moreover, the administration of onionin A_1 suppressed cancer progression and prolonged the survival time in a murine ovarian cancer model under single and combined treatment conditions (Tsuboki et al., 2016).

Recently, onionin A_1 have been reported to be an efective agent for controlling tumors, and that the antitumor effects observed *in vivo* are likely caused by reversing the antitumor immune system. Activation of the antitumor immune system by onionin A_1 might be an efective adjuvant therapy for patients with osteosarcoma, ovarian cancer and other malignant tumors (Nohara et al., 2021).

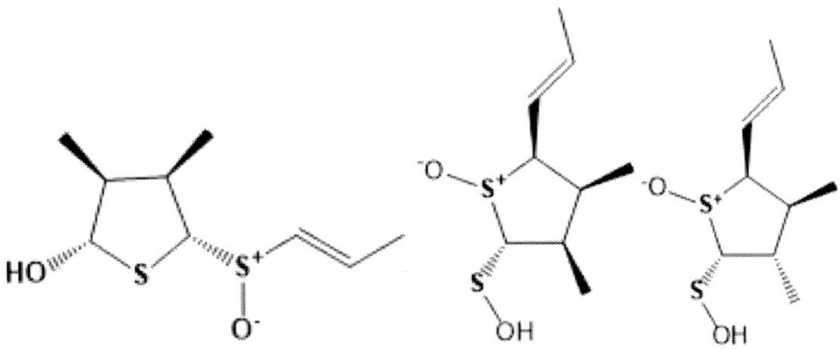

Onionin A_1 Onionin A_2 Onionin A_3

A cyclic sulfide compound of *garlicnin A* was first isolated from acetone extracts of *Allium sativum* L. garlic bulbs and its structure was elucidated as 3,4-dimethyl-5-(4,5-dithia-1*E*,7-octadiene)-tetrahydrothiophene-2-sulfoxide-

S-oxide (El-Aasr et al., 2011). Moreover, garlicnin A displayed the potency to supress tumor cell proliferation by blocking the M2-macrophages polarization (El-Aasr et al., 2011).

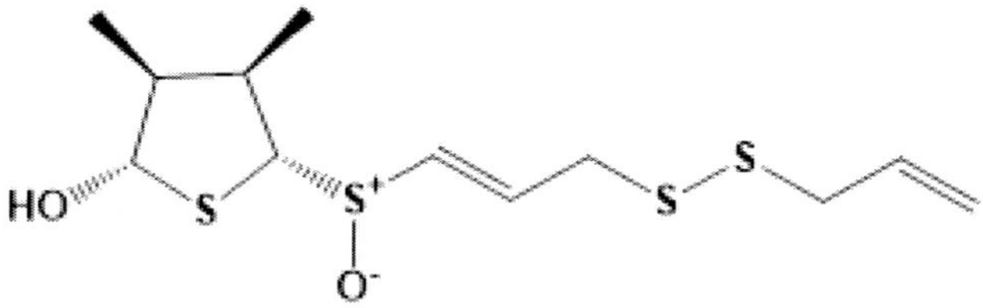

Garlicnins A

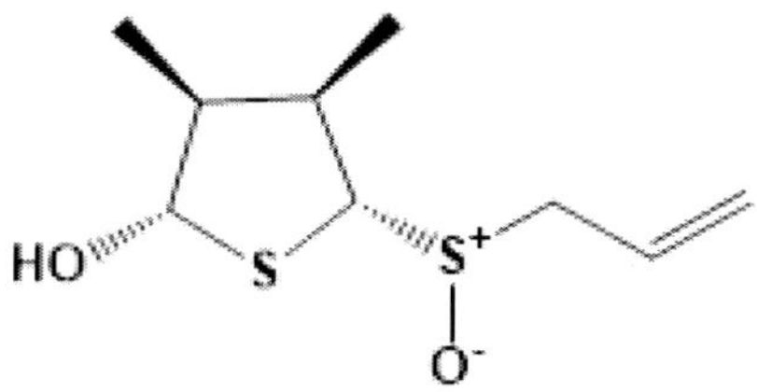

Garlicnins B_1

Garlicnin B_1 was first isolated as 2-sulfenic acid-5-allyl-3,4-dimethyltetrahydrothiophene-2-sulfoxide-*S*-oxide by Nohara's group in 2012 year (Nohara et al., 2012). However, due to the continuity of 9 carbons was not observed in 2D ^{13}C-^{13}C NMR incredible natural abundance double quantum transfer experiments (INADEQUATE), garlinin B_1's chemical structure was revised as 3,4-dimethyl-5-allylsulfinylthiolane-2-ol and proposed as an ajothiolane by Block et al., 2018. Garlicrins B_2, B_3, and B_4 were characterized stereoisomers of B_1 which is the most abundant garlic sulfide.

In a recent study, garlicnin B_1 was unexpectedly reported to display potent anti-inflammatory against DSS-induced mouse colitis in not-dose dependent manner. Furthermore, a significant increase in the levels of inflammatory cytokines, such as MCP-1 and TNF-α in DSS-applied mice were observed. Additionally, garlicnin B_1 was reported to exhibit strong cytotoxicity activity against C26 colon cancer cells and it was able to eliminate all cancer cells at a concentration of 100 μg/mL (Gao et al., 2023).

Garlicnin C_1 was isolated from acetone extracts of bulbs of *Allium sativum* L. Its stereoisomers C_2 and C_3 were isolated from Allium species by Nohara and colleagues (Nohara et al., 2013).

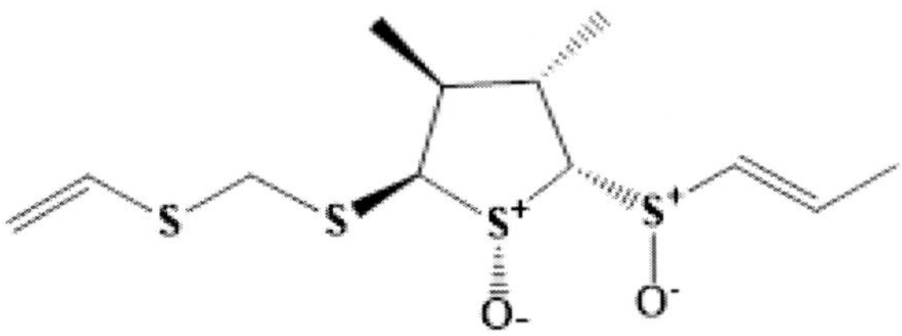

Garlicnins C_1, C_2, C_3

Conclusion

According to National Cancer Institute, *Allium sativum* L. (garlic) is ranked at the top of list of designer foods showing anticancer effects (Theisen 2001). The use of garlic and garlic-based extracts has been linked to decreased incidence of cancer in epidemiological studies. Anticancer activity of garlic based on carcinogen metabolism through inducing cellular gluthathione synthesis which triggers cell cycle arrest and apoptosis pathways, and also inhibition of *Helicobacter pylori* infection which is main cause of gastric, and colorectal cancers (Gao et al., 2023; Smith 2010; Rose et al., 2005). Garlicnin B_1 has been recent expected to be design as a new anticancer drug, due to its stability, readily isolation and chemically synthesizability (Nohara et al., 2021). On the other hand, certain cyclic sulfur-containing compounds might not be promising agents, because of the presence of highly reactive sulfenic acid and thiolactone sulfoxide moieties.

So far, numerous acyclic sulfite compounds have been isolated from *Allium* plant species. It is not possible at this stage to discuss a pharmacological effect difference between cyclic and acyclic sulfide compounds, since the structure-function relationship of the compounds has not been fully elucidated.

Ackowledgment

The author would like to thank of the Vice Chancellor for Research of Freie Universitat Berlin for their approval and financial support of this study.

Conflict of Interest Statement

The author declares no conflict of interest.

References

Cavallito C J, Bailey J H (1944) Allicin, the antibacterial principle of Allium sativum. I. Isolation, physical properties and antibacterial action. *J. Am. Chem. Soc.* 66:1950–1951.

Dale, J. A.; Mosher, H. S. Nuclear magnetic resonance enantiomer regents. Configurational correlations via nuclear magnetic resonance chemical shifts of diastereomeric mandelate, O-methylmandelate, and alpha.methoxy.alpha.-trifluoromethylphenyl acetate (MTPA) esters. *J. Am. Chem. Soc.* 1973, 95, 512–519.

El-Aasr M, Fujiwara Y, Takeya M, Ikeda T, Tsukamoto S, Ono M, Nakano D, Okawa M, Kinjo Ji Yoshimitsu H, Nohara T. 2010. Onion a from Allium cepa inhibits macrophage activation. *Journal of Natural Products.* 73,7:1306-1308.

Gao S, Yang K, Nohara T, Ikeda T, Zhou J-R, Yokomizo K, Fang J. Garlicnin B1, an Active Cyclic Sulfide from Garlic, Exhibits Potent Anti-Inflammatory and Anti-Tumor Activities. *Antioxidants.* 2023; 12(4):869. https://doi.org/10.3390/antiox12040869.

Gao S H, Nohara T, Ikeda T, Zhou J R, Yokomizo K. 2021. Novel acyclic sulfide, garlicrin L-5, from garlic. *Current Topics in Phytochemistry.* 17:29-34.

Gao S, Yang K, Nohara T, Ikeda T, Zhou J-R, Yokomizo K, Fang J. Garlicnin B1, an Active Cyclic Sulfide from Garlic, Exhibits Potent Anti-Inflammatory and Anti-Tumor Activities. *Antioxidants.* 2023; 12(4):869. https://doi.org/10.3390/antiox12040869.

Nohara T, El-Aasr M, Ikeda T, Goa SH, Yokomizo K. 2021. Determination of absolute configuration of the most abundant garlic sulfide, garlicnin B_1. *Current Topics in Phytochemistry.* 17: 55-61.

Nohara T, Kiyota Y, Sakamoto T, Hideyuki Manabe, Masateru Ono, Tsuyoshi Ikeda, Yukio Fujiwara, Daisuke Nakano, and Junei Kinjo. 2012. Garlicnins B1, C1 and D, from fraction regulating macrophage activation of *Allium sativum. Chem. Pharm. Bull.* 60(6):741-751.

Nohara T, Fujiwara Y, Kudo R, Yamaguchi K, Ikeda T, Murakami K, Ono M, Kajimoto T, Takeya M. 2014. Isolation and characterization of new onionins A_2 and A_3 from *Allium cepa*, and of Onionins A_1, A_2, and A_3 from *Allium fistulosum. Chem. Pharm. Bull.* 62 (11):1141-1145.

Nohara, T., Fujiwara, Y., El-Aasr, M. Tsuyoshi Ikeda, Masateru Ono, Daisuke Nakano & Junei Kinjo. Thiolane-type sulfides from garlic, onion, and Welsh onion. *J. Nat. Med.* 75, 741–751 (2021). https://doi.org/10.1007/s11418-021-01533-x.

Nohara T, Fujiwara Y, El-Aasr M, Ikeda T, Ono M, Nakano D, Kinjo J. 2021. Thiolane-type sulfides from garlic, onion and Welsh onion. *Journal of Natural Medicines* 75:741-751.

Rose P, Whiteman M, Moore P K, Zhu Y Z (2005) Bioactive *S*-alk(en)yl cysteine sulfoxide metabolites in the genus *Allium*: the chemistry of potential therapeutic agents. *Nat. Prod. Rep.* 22:351–368.

Stafano I, Zapal J, Moos M, Kuzma M, Kubec R. 2019. Isoalliin-Derived Thiolanes Formed in Homogenized Onion. Journal of Agricultal and Food Chemistry. 67(35):9895-9906.

Smith S (2010) Chemistry in a salad bowl: *Allium* chemistry and biochemistry. In: Block E (ed) Garlic and other alliums. Royal Society of Chemistry; RSC Publishing, *The lore and science*, pp100–223.

Sullivan, G. R.; Dale, J. A.; Mosher, H. S. Correlation of configuration and fluorine-19 chemical shifts of.alpha.-methoxy-.alpha.-trifluoromethylphenyl acetate derivatives. *J. Org. Chem.* 1973, 38, 2143–2147.

Stoll A, Seebeck F (1951) Die synthese des natuerlichen alliins und seiner drei optisch aktiven isomeren. 5. Mitteilung ueber substanzen Helv Chim Acta 34:481–487.

Theisen C (2001) Whatever happened to… ? Looking back 10 years. *J. Nat. Cancer Inst.* 93(14):1049–1050.

Tsuboki, J., Fujiwara, Y., Horlad, H. Daisuke Shiraishi, Toshihiro Nohara, Shingo Tayama, Takeshi Motohara, Yoichi Saito, Tsuyoshi Ikeda, Kiyomi Takaishi, Hironori Tashiro, Yukihiro Yonemoto, Hidetaka Katabuchi, Motohiro Takeya & Yoshihiro Komohara. Onion in A inhibits ovarian cancer progression by suppressing cancer cell proliferation and the protumour function of macrophages. *Sci. Rep.* 6, 29588 (2016). https://doi.org/10.1038/srep29588.

Yoshida, M.; Kameyama, M.; Hosoda, H.; Shimizu, Y.; Sakagami, K.; Washino, K.; Iwata, J.; Omoto, M. *New thiolane compound and use thereof*; JP143866., 2010. https://www.fao.org/faostat/en FAO Food and Agriculture Organization STAT.

Index

T

V